BISHOP BURTON LRC
WITHDRAWN

Accession Number T036339

Class Number 333.72 Ref

Countryside Stewardship: Farmers, Policies and Markets

Related journals

Agricultural Economics
Editor-in-Chief: S.R. Johnson

Food Policy
Editor: K. Salahi

Journal of Rural Studies
Editor: P. Cloke

Land Use Policy
Editor: A. Mather

Free specimen copies of journals available on request

This book presents results obtained within the EU project FAiR1/CT95/0709 on market effects of Countryside Stewardship Policies. It does not necessarily reflect the view of the EU and in no way anticipates the Commission's future policy in this area.

Countryside Stewardship: Farmers, Policies and Markets

Guido Van Huylenbroeck

Faculty of Agricultural and Applied Biological Sciences,
Department of Agricultural Economics,
University of Gent, Belgium

Martin Whitby

Centre for Rural Economy,
Department of Agricultural Economics and Food Marketing,
University of Newcastle–upon–Tyne, UK

1999

PERGAMON
AMSTERDAM – LAUSANNE – NEW YORK – OXFORD – SHANNON – SINGAPORE – TOKYO

ELSEVIER SCIENCE Ltd
The Boulevard, Langford Lane
Kidlington, Oxford OX5 1GB, UK

© 1999 Elsevier Science Ltd. All rights reserved.

This work is protected under copyright by Elsevier Science, and the following terms and conditions apply to its use:

Photocopying
Single photocopies of single chapters may be made for personal use as allowed by national copyright laws. Permission of the Publisher and payment of a fee is required for all other photocopying, including multiple or systematic copying, copying for advertising or promotional purposes, resale, and all forms of document delivery. Special rates are available for educational institutions that wish to make photocopies for non-profit educational classroom use.

Permissions may be sought directly from Elsevier Science Rights & Permissions Department, PO Box 800, Oxford OX5 1DX, UK; phone: (+44) 1865 843830, fax: (+44) 1865 853333, e-mail: permissions @elsevier.co.uk. You may also contact Rights & Permissions directly through Elsevier's home page (http://www.elsevier.nl), selecting first 'Customer Support', then 'General Information', then 'Permissions Query Form'.

In the USA, users may clear permissions and make payments through the Copyright Clearance Center, Inc., 222 Rosewood Drive, Danvers, MA 01923, USA; phone: (978) 7508400, fax: (978) 7504744, and in the UK through the Copyright Licensing Agency Rapid Clearance Service (CLARCS), 90 Tottenham Court Road, London W1P 0LP, UK; phone: (+44) 171 631 5555; fax: (+44) 171 631 5500. Other countries may have a local reprographic rights agency for payments.

Derivative Works
Tables of contents may be reproduced for internal circulation, but permission of Elsevier Science is required for external resale or distribution of such material.
Permission of the Publisher is required for all other derivative works, including compilations and translations.

Electronic Storage or Usage
Permission of the Publisher is required to store or use electronically any material contained in this work, including any chapter or part of a chapter.

Except as outlined above, no part of this work may be reproduced, stored in a retrieval system or transmitted in any form or by any means, electronic, mechanical, photocopying, recording or otherwise, without prior written permission of the Publisher.
Address permissions requests to: Elsevier Science Rights & Permissions Department, at the mail, fax and e-mail addresses noted above.

Notice
No responsibility is assumed by the Publisher for any injury and/or damage to persons or property as a matter of products liability, negligence or otherwise, or from any use or operation of any methods, products, instructions or ideas contained in the material herein. Because of rapid advances in the medical sciences, in particular, independent verification of diagnoses and drug dosages should be made.

First Edition 1999

Library of Congress Cataloging-in-Publication Data
Huylenbroeck, Guido van.
    Countryside stewardship : farmers, policies, and markets / by Guido Van Huylenbroeck,
  Martin Whitby,
      p. cm.
"This book presents the results of the shared cost research project "Market effects of countryside stewardship policies", also called STEWPOL-project, which started in April, 1996 ... [and includes] data in eight EU Member States"– P..
    ISBN 0-08-043587-4 (hc)
     1. Agriculture and state – European Union countries – Case studies. 2. Agriculture – Economic aspects – European Union countries – Case studies. 3. Agriculture – Environmental aspects – European Union countries – Case studies. I. Whitby, Martin Charles. II. Title.
HD1918.H89 1999
338.1'84–dc21   99-045052

British Library Cataloguing in Publication Data
A catalogue record from the British Library has been applied for.

ISBN: 0 08 043587 4

∞ The paper used in this publication meets the requirements of ANSI/NISO Z39.48-1992 (Permanence of Paper).

Printed in The Netherlands.

G. Van Huylenbroeck and M. Whitby
Countryside Stewardship: Farmers, Policies and Markets
© 1999 Elsevier Science Ltd. All rights reserved

Editors' preface

This book presents the results of the shared cost research project "Market effects of countryside stewardship policies", also called STEWPOL-project, which started in April 1996. At that moment most countries were still in the development stage of agri-environmental programmes. Little was known about the effects of such policies. It was therefore a real challenge for all partners to start this research and to collect comparable data in eight EU Member States. The collection and analysis of original empirical data on the application of this new area of policy is probably one of the main achievements of the project besides its new theoretical and methodological findings.

The production of volumes such as this impose costs on many people who do not feature on the contents page. In this case, there are so many contributors that it is impossible to mention them all in detail.

In the first place, the researchers are especially grateful to European taxpayers who, through their representatives in Brussels, have funded this project in the FAIR framework as project FAIR1/CT95-0709. The value to STEWPOL participants, in addition to the professional interest of the work, has been profound. We can also confirm beyond doubt that the social benefits of cohesion generated through such work can be very real. We started as colleagues in 1996, but ended as friends who appreciate each other not only as professionals.

We also recognise a substantial debt to the 2,000 farmers and many officials who answered lengthy questionnaires and provided us with all necessary information for this project, providing a unique cross-section of data from eight countries. That material has informed more than one chapter in this book and adds an important note of veracity to its main conclusions.

A special word of thank goes to Alain Coppens and Marijke D'Haese who helped us in the final editing process, as well as to all others who in one way or another have contributed scientifically or technically to this book. Finally, most STEWPOL participants, and certainly the editors, owe substantial debt to their partners and families who have endured their absorption into this project.

Guido Van Huylenbroeck Martin Whitby
University of Gent *University of Newcastle-upon-Tyne*

G. Van Huylenbroeck and M. Whitby
Countryside Stewardship: Farmers, Policies and Markets
© 1999 Elsevier Science Ltd. All rights reserved

List of contributors to the book and project

Austria
Marcus Hofreither
Franz Sinabell
Klaus Salhofer

Institut für Wirtschaft, Politik und Recht
Universität für Bodenkultur
Gregor Mendel Strasse 33
1180 Vienna
Austria

Belgium
Guido Van Huylenbroeck
Alain Coppens
Evelyn Goemaere
Isabel Vanslembrouck

Vakgroep Landbouweconomie
Universiteit Gent
Coupure Links 653
9000 Gent
Belgium

France
François Bonnieux
Pierre Dupraz
Jean-Paul Fouet

Institut National de la Recherche Agronomique
Unité d'Economie et de Sociologie de Rennes
Rue Adolphe Bolierre
CS 61103
35011 Rennes Cedex
France

Germany
Stephan Dabbert
Ottmar Röhm

Institut für Landwirtschaftliche Betriebslehre
Universität Hohenheim
Schloss Osthof Süd
70599 Stuttgart
Germany

Greece
Dimitris Damianos
Yannis Barlas
Efthalia Dimara

Department of Economics
University of Patras
University Campos-Rio
P.O. Box 1391
26500 Patras
Greece

Italy
Maurizio Merlo
Paola Gatto

Dipartemento Territorio e Sistemi Agro-Forestali
Sezione Economia
Universita degli Studi di Padova
Agripolis
Via Romea
35020 Legnaro
Italy

Sweden
Lars Drake
Per Bergström
Henrik Svedsäter

Department of Economics
Swedish University of Agricultural Sciences
P.O Box 7013
75007 Uppsala
Sweden

United Kingdom
Martin Whitby
Katherine Falconer
Christopher Ray

Centre for Rural Economy
University of Newcastle-upon-Tyne
6, Kensington Terrace
Newcastle-upon-Tyne NE1 7RU
United Kingdom

List of abbreviations

AEP	Agri-Environmental Programme
Agenda2000	Reform programme of the Common Agricultural Policy for the year 2000
AOC	Appellation d'origine controllée
CAP	Common Agricultural Policy
CS	Countryside Stewardship
CSP	Countryside Stewardship Policy
EBD	Environmental Bad and Dis-service
EDM	Equilibrium Displacement Model
EGS	Environmental Good and Service
ERBD	Environmental and Recreational Bad and Dis-service
ERGS	Environmental and Recreational Good and Service
ESA	Environmental Sensitive Area
EU	European Union
F&F	Food and Fibre
LFA	Less Favoured Area
LU	Livestock Unit
OECD	Organisation for Economic Co-operation and Development
PMP	Positive Mathematical Programming
PPC	Production Possibility Curve
PPP	Polluter Pays Principle
PQP	Positive Quadratic Programming
PSE	Producer Support Estimate
STEWPOL	STEWardship POLicies research project (FAIR1/CT95/0709)
UAA	Unit of Agricultural Area

Exchange rate for currencies

Currency	Country	1 EURO or ECU =
ATS	Austria	13.760 ATS
BEF	Belgium	40.339 BEF
FRF	France	6.559 FRF
ITL	Italy	1,936.270 ITL
DEM	Germany	1.956 DEM
GRD	Greece	324.037 GRD
GBP	United Kingdom	0.661 GBP
US$	United States	1.050 US$

G. Van Huylenbroeck and M. Whitby
Countryside Stewardship: Farmers, Policies and Markets
© 1999 Elsevier Science Ltd.   All rights reserved

Contents

Editors' preface		v
List of contributors to the book and project		vii
List of abbreviations and exchange rates		ix
1.	**Introduction to research on Countryside Stewardship Policies**	1
1.1.	Introduction	1
1.2.	The STEWPOL-project	4
1.3.	The European policy context	6
	1.3.1. European environmental policies	6
	1.3.2. Rural development policies	7
	1.3.3. Agri-environmental policies	7
	1.3.4. Product quality policies	8
1.4.	The national context in the partner countries	8
	1.4.1. Economic structure of agriculture in STEWPOL countries	9
	1.4.2. Impact of agriculture on the environment	10
	1.4.2.1. Land and soil degradation	11
	1.4.2.2. Water resources	11
	1.4.2.3. Impact on landscape and bio-diversity	12
	1.4.3. Institutional framework for agri-environmental policies	15
	1.4.3.1. Administrative structure	15
	1.4.3.2. Implementation of Regulation 2078/92	15
1.5.	Outline of the book	16
1.6.	Conclusions	18
2.	**The economic nature of stewardship: complementarity and trade-offs with food and fibre production**	21
2.1.	The rationale of Countryside Stewardship Policies (CSPs)	21
	2.1.1. What is Countryside Stewardship?	21
	2.1.2. The Production Possibility Curve (PPC) between Food and Fibre (F&F), Environmental Recreational Goods and Services (ERGSs) and Environmental Recreational Bads and Dis-services (ERBDs)	23
	2.1.3. Boundaries and reference points between ERGSs and ERBDs	26
2.2.	Options and tools for implementing Countryside Stewardship Policies	28
	2.2.1. Mandatory regulations	28
	2.2.2. Financial-economic voluntary tools	30
	2.2.3. Market creation	31
2.3.	The survey of CSPs	32

		2.3.1.	Methodology	32
		2.3.2.	Objectives of CSPs	34
		2.3.3.	Rationale and attributes of CSPs	35
		2.3.4.	Administration	37
		2.3.5.	Target areas, uptake and average remuneration	38
		2.3.6.	Commodities and markets potentially affected by CSPs	40
		2.3.7.	Farm practices affected by CSPs: trade-offs between F&F, ERGSs and ERBDs	41
	2.4.	Conclusions		42
3.	**Policy indicators and a typology of instruments**			**47**
3.1.	Introduction			47
3.2	Definition and selection of criteria			49
	3.2.1.	General guidelines		49
	3.2.2.	Overview of criteria		51
		3.2.2.1.	Environmental effectiveness	51
		3.2.2.2.	Economic efficiency	52
		3.2.2.3.	Sustainability	52
		3.2.2.4.	Political acceptability	52
	3.2.3.	Analysis of criteria of special interest		53
3.3.	Establishing a typology of the CSPs			55
	3.3.1.	Principles of the methodology		55
	3.3.2.	Factor selection		58
3.4.	Graphic representation of the typology			59
	3.4.1.	Definition of homogeneous groups of CSPs		59
	3.4.2.	Representation of G1, G2 and G7 on the "output map"		60
	3.4.3.	Representation of G3, G4 and G5 on the "input map"		61
	3.4.4.	Representation of G6 on an "output-input" map		63
3.5	Concluding comments			64
4.	**The invisible costs of scheme implementation and administration**			**67**
4.1.	Introduction			67
4.2.	Transactions costs and contracting in agri-environmental policy			69
4.3.	Empirical administration cost estimations for agri-environmental schemes			71
	4.3.1.	Cost typologies and incidence		71
	4.3.2.	Methodology		73
4.4.	Findings from the case studies			76
4.5.	Discussion			82
	4.5.1.	General factors affecting administrative cost levels		82
	4.5.2.	Where does the burden of transactional costs fall?		84
	4.5.3.	What scope is there to reduce the administrative costs of agri-environmental schemes?		84
	4.5.4.	Scheme value for money		86
4.6.	Conclusions			87
5.	**Farmers' attitudes and uptake**			**89**
5.1.	Introduction			89
5.2.	Theoretical issues			90
	5.2.1.	Model for farm behaviour		90
	5.2.2.	Attitude measurement		92
5.3.	Methodology and data collection			93
	5.3.1.	Pre-test		93
	5.3.2.	Questionnaire design		93
	5.3.3.	Data collection		94
	5.3.4.	Regression models		96
	5.3.5.	Hypotheses		96
5.4.	Analysis of results			97
	5.4.1.	Some general results		97
	5.4.2.	Model explaining attitude		98
	5.4.3.	Model explaining participation		103
5.5.	Summary and conclusions			110

6.	**Modelling regional production and income effects**	113
6.1.	Introduction	113
6.2.	The methodological approach	114
	6.2.1. Choosing the policies to be investigated	114
	6.2.2. Choosing the appropriate type of model	115
	6.2.3. Positive mathematical programming	116
6.3.	Case studies of 8 European regions	120
6.4.	Oberösterreich—detailed modelling results for a European region that is a leader in CSPs	122
	6.4.1. Geography and climate, agricultural production and CSPs	122
	6.4.2. The structure of the regional agricultural sector model of Oberösterreich	124
	6.4.3. Effects of CSP implemented in Oberösterreich	124
	6.4.4. Effects on farm income and profiles, dual values and budgets	125
	6.4.5. Cropping structure, structure of production methods and output effects	127
6.5.	The effects of CSPs in the other study regions	130
6.6	CAP, Agenda 2000 and their relevance for CSPs	132
7.	**Estimating aggregate output effects**	135
7.1.	Introduction	135
7.2.	Classification of CSPs according to the instruments that are used	137
7.3.	The role of property rights and direct payments	140
7.4.	Methods to evaluate market effects	141
7.5.	Results on the output consequences of CSPs from a farm survey	143
7.6.	Quantitative output effects of countryside stewardship schemes	144
	7.6.1. The method of equilibrium displacement models	144
	7.6.2. The Austrian 'crop rotation scheme'	147
	7.6.3. Parameters of the equilibrium displacement model	148
	7.6.4. Quantitative results of the equilibrium displacement model	150
7.7.	Summary and conclusions	152
8.	**Analysis of trade distortions**	157
8.1.	Introduction	157
8.2.	The relationship between agri-environmental policies and trade: theoretical considerations	158
	8.2.1. General framework	158
	8.2.2. Trade effects of agri-environmental policies: theoretical considerations	159
	8.2.3. Evidence	160
8.3.	The agri-environmental aspects of the Uruguay Round Agreement	161
	8.3.1. Main elements and implications	161
8.4.	Methodology for measuring trade distortion	164
	8.4.1. Alternative approaches	164
	8.4.2. The Producer Support Estimate (PSE)	165
	8.4.3. The qualitative approach	166
8.5.	Results of the quantitative and qualitative analysis	168
	8.5.1. Quantitative analysis	168
	8.5.2. Qualitative analysis	168
	8.5.2.1. Comparative statistical analysis	171
	8.5.2.2. Cluster analysis	172
8.6.	Discussion and conclusions	175
9.	**Conclusions and policy recommendations**	177
9.1.	Introduction	177
9.2.	Main outcomes of the research	178
	9.2.1. Policies	178
	9.2.2. Farmers	179
	9.2.3. Market effects	179
9.3.	Discussion	180
	9.3.1. Are countryside stewardship policies an efficient way to achieve countryside management objectives?	180

	9.3.2.	Improving the outcome of policies		181
		9.3.2.1. Targeting of policies		181
		9.3.2.2. Increasing uptake		182
		9.3.2.3. Policy administration and mechanisms		183
	9.3.3.	The transaction costs of policies		184
	9.3.4.	Market and trade effects		186
	9.3.5.	Political acceptability		187
9.4.	Conclusions			188

Annexe: Overview of inventoried policies 193
Austria 194
Belgium 197
France 203
Germany 205
Greece 215
Italy 217
Sweden 225
United Kingdom 227

Index 229

G. Van Huylenbroeck and M. Whitby
Countryside Stewardship: Farmers, Policies and Markets
© 1999 Elsevier Science Ltd. All rights reserved

Chapter 1

Introduction to research on Countryside Stewardship Policies

Guido Van Huylenbroeck, Alain Coppens and Martin Whitby

Abstract—In this opening chapter, the focus of the book is presented, starting with an introduction to the problem of countryside stewardship, the reasons why it has been neglected by modern agriculture and why government intervention is necessary to provide the public goods desired by society from agriculture. In the second section the STEWPOL project, a three year research project on the actual implementation of agri-environmental measures or countryside stewardship policies in eight participating countries, is presented as well as the links between the different parts of the research. Before continuing with an outline of the book, the European policy context and the agrarian structure in the eight STEWPOL countries are summarised and the environmental problems and the institutional framework in which agri-environmental policies operate in these countries are reviewed.

1.1. Introduction

The loss of environmental quality and of rural landscapes is one of the major contemporary challenges for developed countries. Agriculture and forestry, occupying most of the rural land throughout Europe, are the activities that have a major influence on the European rural environment and landscape. The European countryside is mainly man-made, resulting from centuries of management of rural areas for food and fibre (F&F) production. Until the last few decades, this has caused few problems because agriculture was traditionally a closed system, where the land was used for the production of human food and feed for animals, producing manure to be returned to the land as source of nutrients. It was also in the interest of the rural population and landowners to maintain or improve the resource base of the land by carrying out stewardship practices such as the maintenance of roads and hedges, drainage and water regulation systems, woodland management and so on (Ferro *et al.*, 1995). These practices also brought the advantage of providing, as an unintended by-product, a diverse array of landscapes and environmental goods and services (EGSs). These public goods are now regarded as a key environmental asset and are highly valued.

As long as agricultural operations were relatively small in scale, before mechanisation and the use of external inputs, there was a degree of harmony between agriculture and the countryside in which it operated. It was only with the increased use of external inputs (compound feed, chemical fertilisers, pesticides and mechanisation) that the closure of the system was broken, resulting in a decline of the quality and quantity of the EGSs produced. Modern agricultural systems based on a large amount of external inputs provoked an intensification of agriculture in those areas well suited to agricultural activities with the potential for what Gatto and Merlo in Chapter 2 in this book call EBDs (Environmental Bads and Dis-services) or pollution. At the same time, intensification in these regions leads to land abandonment in other regions less suitable for agriculture, producing further negative effects on the quality of the countryside.

As countries have become more affluent, the demand for environmental goods has gained increasing importance as an item of policy. This reflects the shift in the emphasis of consumer concerns from that of securing an adequate food supply towards taking a greater interest in environmental goods. In economic terms it is reasonable to argue that the environment is an income elastic good, which means that it will become more important to citizens as incomes rise, in contrast with raw food, of which stable populations demand very little more as they become richer. Unfortunately we do not have reliable measures of the relevant elasticities, although for food at the farm gate it is generally agreed to be low, even negative, for some foods. The estimation of aggregate elasticities of demand with respect to environmental goods has not yet been achieved and is difficult as in most cases consumers do not have to pay for their delivery (in economic terms they are external to the market).

In this situation a consideration of, and reflection on, the objectives and impact of agricultural practices in terms of countryside management and stewardship is necessary. Newby *et al.* (1978) have pointed out that countryside stewardship was adopted as a kind of philosophy or duty of the landed gentry of England in the eighteenth and nineteenth centuries, whilst they also maintained the productive base of the land they owned. Colman (1994) has examined the economic questions raised by the phenomenon of stewardship in a neo-classical framework, underlining the dilemma for policy makers that, by offering rewards for stewardship practices, they risk undermining farmers' and land users' ethical commitment to the continued supply of such goods where there is no reward. The stewardship role persists but has decreased and there is often a lack of countryside stewardship, especially where relevant practices are no longer needed or are less necessary for modern agriculture, leading to the degradation of landscapes and of the natural resource base.

As society has become aware of this degradation, the countryside stewardship and management role of agriculture is increasingly seen in terms of the maintenance or restoration of remaining natural, environmental and scenic resources. Governments have introduced policies not only to persuade farmers, as the custodians of rural resources, to contribute positively to the production of EGS but also to avert further degradation. This intervention of society is necessary as normal market mechanisms fail to regulate this production. The absence of individual property rights make it impossible to establish a private market in which prices (market valuations) guide resource allocations. Most farmers will not contribute to EGS-production as long as this only results in higher costs for them (as the required practices are no longer a necessary part of their productive activity) rather than in higher remuneration. It is then the role of policy to design and introduce systems that make farmers aware of their countryside stewardship function and deliver incentives which will persuade them to change their practices.

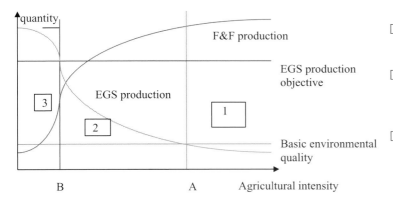

Fig. 1.1. **Trade-off between F&F and EGS-production in rural areas and link with policy instruments for countryside stewardship**.

The objective of countryside stewardship policies is thus to reconcile the balance between F&F production on the one hand and EGS production on the other and to ensure the availability of appropriate quantities of both. In general this can be achieved through two major mechanisms. The first possibility is regulation, the second is through economic mechanisms. Regulation will mainly be used to avoid unacceptable levels of negative effects, while economic mechanisms will be used to stimulate contributions of farmers above their legal obligation. Figure 1.1 schematises this difference by depicting the trade-off between F&F production and EGS-provision. Point A reflects the minimum environmental quality that society wants to be respected. This means that all forms of agricultural activity situated in zone 1 make such a use of their private property rights that it damages the individual property rights of others on, e.g. pure water or clean air. In a democracy, the location of this point is a question for public debate. This point also shifts in time because of changing preferences within a society (cf. the above elasticity discussion). When this point is identified, society has the legal right to enforce its recognition through legal and binding instruments. Following the polluter-pays-principle (PPP) the decrease in production or income caused by social limitations does not have to be compensated, although society often provides some form of compensation when the limitation of individual property rights is particularly important. The difficulty of setting reference points is further discussed in Chapter 2.

In zone 2 (between point A and B), economic instruments may be used to persuade producers to implement those practices, which are not enforceable because of property rights but desired by society, in their own interest on a free basis. Theoretically further distinction can be made between zone 2 and zone 3. Where in zone 2 economic instruments changing the price conditions between F&F and EGS have to be implemented, in zone 3 instruments reducing abandonment of land and agricultural production are needed as in this situation agriculture is a necessary condition to maintain the value of the prevailing EGS. This is the case in designated areas with high nature or landscape value. In these zones subsidies to keep farmers on their land may be justified. For a more thorough discussion of these instruments we refer to Chapter 2.

As agriculture is already the focus of massive government intervention both through national and EU policies, further policies aiming to secure environmental gains for society have to fit in with existing polices, indeed often to counter their worst impacts. The agricultural sector is also under the influence of other profound forces for change arising mainly from the industries which have so successfully designed and delivered the new inputs it uses. No study of agriculture and the environment will succeed if it concentrates solely on policies and farmers without recognising these other major external pressures on the system. The interaction between agricultural production and trade policies and those concerned with delivering environmental goods is a crucial one, which this study has addressed.

The remainder of this chapter first describes the research project of which this book is the result, situates the policies analysed within the EU-policy framework and reviews the agri-environmental situation in the countries analysed before providing a brief outline of the book.

1.2. The STEWPOL-project

Given the limited theoretical, and in particular empirical knowledge of the application of agri-environmental policies, the EU-Commission decided to finance a shared-cost three year research project on this issue which started April, 1st 1996. Eight research teams from Austria, Belgium, France, Germany, Greece, Italy and the UK were involved. The central objective of the project was an analysis of the economic nature and market effects of actual applied countryside stewardship policies (CSPs) in the eight participating countries, here called STEWPOL countries. A broad definition of CSPs has been used, namely all policies that aim at guarding—in the sense of managing and preserving—the natural environment and the farmed landscape. The definition has been wide enough to cover not only Regulation 2078/92 measures but also all kinds of national, regional or local policies used in the same area. In total the research was divided into seven tasks: a study of the economic nature of countryside stewardship policies, the construction of an overall typology of the policies applied, an analysis of their administrative costs, the determination of farmers' attitude towards such policies, their effects at farm level and on agricultural commodity markets, their impact on trade distortion and an overall evaluation of the policies. All these tasks are reported in the chapters of this book. The link between these chapters is schematically represented in fig. 1.2.

Two groups can be distinguished: on the one hand a group of tasks that has analysed empirical information on a number of attributes of CSPs applied in the eight STEWPOL countries. They all relate to the inventory of policies, collected during the first year of the project. It concerns Chapter 2 (classification of policies) mainly studying the economic nature of CSPs, Chapter 3 that has supplemented the inventory with a number of data allowing to construct a typology of CSPs based on multiple criteria, Chapter 4 which has collected and analysed extra information on the administrative costs related to the development, implementation, monitoring and control for a sample of the inventoried policies and finally Chapter 8 which has classified the inventoried policies according to criteria reflecting their non-trade distorting character.

The second group of tasks was mainly oriented towards the impact of CSPs at farm and agricultural commodity market level. Chapter 5 has used a survey to measure farmers' attitudes towards CSPs and the reasons for participation. The fieldwork for this survey was done in the first half of 1998. This task has also collected evidence about farmers' opinion on the output reduction effect of CSPs and about the share of received premiums reinvested in the farms, information further used in Chapter 7. Another approach has been followed in analysing the income and output effects. Because of the limited time frame of implementation of most CSPs a normative approach based on mathematical and economic models has been judged more suitable for these tasks. The income and output effects have been modelled at regional level in Chapter 6 using a positive mathematical programming model. Based on the results of this model and the information from the survey, Chapter 7 tries to explain why CSPs in general have only low aggregate market effects, using an equilibrium displacement model.

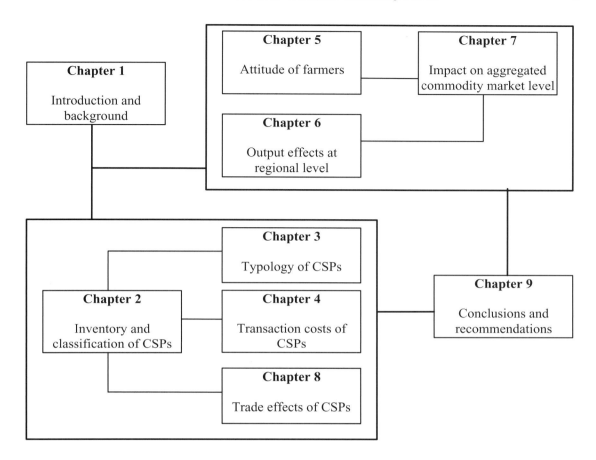

Fig. 1.2. **Structure of the book**.

In the following chapters, the methodologies applied for each task are further described. It is however important to emphasise that all partners have contributed and put considerable efforts in data collection for all tasks. For each of the research tasks, one or more partners—the respective authors of the chapters in this book—were responsible for the analysis, but all partners contributed to data collection for each of the tasks. Most of the data reported in this book are original and have been collected through analysis of legislation, from administrative sources and through interviews with officers and farmers. The fact that these data have been collected in the same way in eight different countries gives them an added comparative value and increases the interest of the empirical observations. It is important to say that although efforts have been made to avoid judgements, they could not be totally excluded as for many aspects of this project no data were available. However as no systematic bias has been observed, we are confident that within their limitations, the data sets used could serve for the analyses intended.

A methodological decision which was taken at the proposal stage of the project was that the economic value of benefits would not be measured here. The measurement of the benefits of these policies presents a major task which has, so far, been tackled on a piecemeal basis in some member states although none has yet valued the benefits of all such policies. Such studies continue but still a majority of diverse policies is left for which there are no economic benefit measurements at all. In those circumstances it was decided that benefit measurements would not be considered, nor the issue of pure environmental indicators. This might be considered as a weakness of this book, but by concentrating on the economic nature and effects of CSPs a higher (disciplinary) homogeneity was reached.

Before outlining the content of the book in more detail, the European and national context in which the programmes analysed are situated are briefly described. At the European level the relation with a number of regulations and programmes is emphasised while the national context stresses the differences in agricultural structure, environmental problems and legislative framework influencing the different application of CSPs between the eight STEWPOL countries.

1.3. The European policy context

It is clear that the agricultural and environmental situation in EU member states is strongly influenced by the CAP. Although it is difficult to assess the precise influence of the CAP on the environment and countryside (Brouwer and Van Berkum, 1996) because of the problem of knowing how agriculture in the different member states would have developed without the CAP, there is no doubt that the CAP has had a serious impact on the development and intensification of the agrarian sector during the past 40 years. As agriculture usually is the main land user, the CAP has also had major impacts on the European countryside.

However, it is only recently that this impact has been recognised at European level and that common policies have been developed emphasising countryside management and stewardship. In this context, four rather independent lines of action can be distinguished:

- European environmental policies influencing agriculture
- Rural development policies within the CAP
- Agri-environmental policies within the CAP
- Food quality policies

1.3.1. European environmental policies

All policies issued by the council of Environmental Ministers on proposal of the EU Commission, and which have as their main objectives the protection of the environment and prevention of pollution, belong to the first category of important policies. With respect to the countryside and agriculture two types of regulations are of importance:

- those defining maximum levels of pollutants such as the Nitrate Directive limiting the N-content to 50 mg/l of groundwater. Although these regulations are not specific to agriculture or countryside, they have a strong impact on it as agriculture is a main source of nitrate pollution in most countries, in particular in those with a high livestock density. Member states have to take measures to decrease N-use in the agrarian sector and this will have a positive impact on the status of the rural environment. In terms of fig. 1.1, such regulations fix the position of point A at an European level.
- those having a strong impact on the countryside management and stewardship role of agriculture, are the assignment directives at European level of regions with a high natural or environmental value like the Birds Directive or the RAMSAR convention for the protection of European wetlands, and so on. These directives, which in terms of fig. 1.1 have more to do with location of point B for certain zones, seek to protect areas with high natural or environ-

mental value. Also here agriculture is not directly in focus, but it is obvious that when the protection of these areas also requires special management programmes, agriculture in these regions will be affected by it.

1.3.2. Rural development policies

The second category of policies indirectly influencing the countryside stewardship function of agriculture are the European rural development programmes better known under the name of objective 1, 2, 5a and 5b (and other labels proposed in Agenda 2000) programmes with initiatives such as LEADER and other programmes. The main objective of these programmes is to co-finance actions of national member states for the development of regions with prescribed deficiencies in economic development or with certain natural handicaps (such as mountainous regions). In regions falling under one of these objectives, special programmes are possible. In particular in the objective 1 and 5b regions, under Agenda 2000 called objective 1 and 2 (rural regions with a lack in agricultural development), these programmes often include actions trying to develop new sources of income for farmers through countryside stewardship activities (for example, the creation of regional labels). Such programmes may also be important for regions with natural handicaps because they provide funds to maintain a certain level of agricultural activity in these regions, so protecting the typical rural landscapes and ecosystems.

1.3.3. Agri-environmental policies

A third category of policies, the one which are the particular focus in this book, are the so called agri-environmental measures. Originally these policies started modestly with Regulation 797/85, under which Article 19 permitted member states to pay farmers in environmentally sensitive areas in return for their adherence to traditional practices. Regulation 1760/87 provided some opportunities of co-financing (up to 25 per cent of compensation) for actions aiming at stimulating positive contributions of the agricultural sector to the conservation of the environment. They have become more important under Regulation 2078/92 (co-financing up to 50 per cent and even 75 per cent in objective 1 regions) one of the so called accompanying measures of the 1992 CAP-reform. This regulation (amended by Regulation 2772/95 and 746/96) provides European financing of programmes under the following 12 headings:

– organic farming
– reducing entrants
– extensification or non-intensification of crop production
– conversion of arable land to grassland
– grassland (system) maintenance
– livestock reduction per unit of area
– preserving local breeds
– landscape and nature protection
– management of abandoned farm and wood land
– 20 year set aside
– recreational access
– training

Under Agenda 2000 these policies are continued and even reinforced within the so-called second pillar of the reform, the structural policies. Before the end of 1999 Member States have to make proposals for their second generation agri-environmental programmes.

Although, a major objective of the 1992 reforms was to provide an alternative source of income to compensate for the decrease in price and income support expected from the reform, this regulation has become the main instrument for countryside stewardship programmes. As indicated by Brouwer and Lowe (1998) and Deblitz and Plankl (1998), there are important differences between programmes and uptake across Member States, reflecting differences of attitude and resource availability, but also their different dynamics. For a comprehensive inventory of the programmes applied in the different EU-Member States, see Deblitz and Plankl (1998). One major aim of the research project was to study the differences between policies and to analyse their success or failure. Broadly speaking the programmes within the different member states can be divided into two categories: horizontal measures, which are payments for which all farmers in the state (or region in federal countries) can apply (in terms of fig. 1.1 policies falling between point A and B) and vertical measures or policies targeted to areas designated on certain criteria (indicated by the situation on the left of point B of fig. 1.1).

1.3.4. Product quality policies

The fourth category of European initiatives influencing the countryside stewardship role of agriculture includes the directives and regulations with regard to the quality of agricultural products, in particular those making possible the differentiation of products by their origin or mode of production (Regulations 2081/92 and 2082/92). These regulations give new opportunities to apply the beneficiary-pays-principle for EGS jointly produced with agricultural commodities. Another regulation falling into the same category is Regulation 2092/91 specifying the conditions for organic farming.

One important general observation on these European action programmes is the lack of co-ordination, complementarity or compliance that can be observed. One example of this is that many of the above programmes or policies are vertical policies restricted to certain zones, although under each programme these zones are designated in a different way. There are for example differences between European designated and protected areas, national areas eligible for agri-environmental complementary payments and the areas eligible for rural development support. A better co-ordination at all levels with a certain but unified degree of protection could improve clarity and save on administration and transaction costs (see further in Chapter 4). The proposals of Agenda 2000 where a closer link is made between agri-environmental programmes and rural development actions and where the principle of cross-compliance can be applied, is probably a step in the right direction.

Another problem is the link with the general CAP measures. Where agri-environmental programmes are necessary to counteract negative impacts of the general CAP-support, in general high support prices and direct income support increase the level of compensatory payments necessary to avoid the negative effects of the price distortion. In the concluding chapter, this issue is further developed as it is clear that certain problems cured by agri-environmental programmes only occur because markets and market prices are distorted.

1.4. The national context in the partner countries

Countryside management is by definition closely related to the specific situation in a region or area. It is therefore not surprising that the policies applied are different from country to country as the situation of the countryside, environment, landscape and agriculture varies. Also traditions, political structures, assignments of property rights and other factors which can influence the kind

Table 1.1
Agricultural land use in STEWPOL countries (1995)

Agricultural land use	Austria	Belgium	France	Germany	Greece	Italy	Sweden	UK
Total land (thousand km^2)	84	31	549	367	132	295	460	241
Agricultural land (per cent)	41	57	55	55	68	56	7	71
Wooded area (per cent)	39	20	27	29	24	21	50	10
Agricultural area (1000 hectares)	3,430	1,368	30,056	17,344	3,941	14,685	3,259	18,294
Arable crops (per cent)	41	53	59	68	51	56	84	29
Grassland (per cent)	57	46	36	30	31	26	15	67
Permanent crops (per cent)	2	5	6	1	18	18	1	4

Source: individual country statistical services.

Table 1.2
Per cent share of agricultural output in STEWPOL countries (1995)

Per cent share of agricultural output	Austria	Belgium	France	Germany	Greece	Italy	Sweden	UK
Crop output	34	42	50	39	70	60	28	37
Animal output	39	42	31	32	15	25	35	36
Animal products	27	16	19	29	15	15	37	27

Source: OECD, "Economic accounts for agriculture—1996 Edition".

of policy instrument that will be applied differ between countries. The objective of the research was to analyse and compare the policies applied in the eight STEWPOL countries and to search for determinants explaining the success or failure of these policies in terms of implementation, costs, uptake, impact on the output of agricultural commodities and trade distortion. This requires the national context in which these policies are implemented to be taken into account. Therefore and in order to be able to interpret the research results discussed in the following chapters correctly, this section gives an overview of the situation in relation to agriculture, its impact on the environment and the policy context in the STEWPOL countries.

1.4.1. Economic structure of agriculture in STEWPOL countries

The structure of agriculture is highly diverse between and within countries. A description of the agrarian structure as in Tables 1.1 to 1.4 therefore risks extreme generality. However, from the figures it is clear that France, Germany, Italy and Sweden all have a large area as well as a large agricultural area. The share of agricultural land in the total area is highest in the UK and Greece and lowest in Austria and Sweden, but especially in the latter, where 50 per cent of the country is occupied by forest. Prominent features with regard to the division of the agricultural area are the share of arable land in Sweden, of grassland in the UK and the large area for permanent Mediterranean crops in Italy and Greece. This use of land is also reflected in

Table 1.3
Farm structural characteristics in STEWPOL countries (1995)

Farms structural characteristics	Austria	Belgium	France	Germany	Greece	Italy	Sweden	UK
Number of farms (1000 farms)	259	73	735	588	862[a]	2478	87	234
Average size (hectares)	21	19	41	31	4.3[a]	5.9	37	73
Employment in agriculture (per cent of active population)	7.3	2.5	4.6	3.3	20.4	8.5	3.1	2.3

[a] 1991.
Source: individual country statistical services

Table 1.4
The 1995 Agricultural output in STEWPOL countries at 1990 prices

Aggregate output	Austria	Belgium	France	Germany	Greece	Italy	Sweden	UK
Value in 1990 prices (10^6US$)	4525	7401	50,743	32,615	12,514	41,248	3580	22,079
Agricultural area (1000ha)	3430	1368	30,056	17,344	39,411	14,685	3259	18,294
Value of output per hectare (1000US$/ha)	1.32	5.41	1.69	1.88	0.32	2.81	1.10	1.21

Source: OCDE, "Economic accounts for agriculture—1996 edition"

the major share of crops in the total output of these two countries. The high level of animal outputs in Belgium reflects the intensive character of its agriculture. The differences in average size of the farms and employment levels are also important when discussing agri-environmental measures. Main differences can be derived from the value of agricultural output per hectare which is an indicator of the intensity of agriculture in the different countries.

1.4.2. Impact of agriculture on the environment

The pressure of agricultural practices on the environment can be regrouped in three main categories:

- land degradation by erosion, compaction and soil pollution (heavy metals, pesticides and salination);
- water resource degradation through pollution by nutrient surpluses, micro pollutants (heavy metals and pesticide residues), bacterial agents and through degradation of the water resources by drainage and irrigation;
- landscape and wildlife deterioration through loss of bio-diversity, habitat destruction and loss of landscape features.

Agriculture also contributes to air pollution, but this is not a problem directly targeted by countryside stewardship policies.

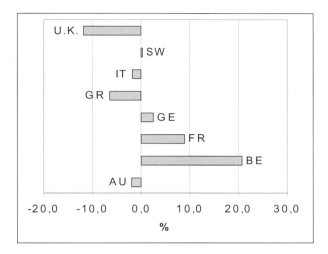

Fig. 1.3. **Changes in the percentage of arable land in total agricultural land use between 1980 and 1995 in the STEWPOL Countries**
Source: OECD (1997)

Also for these aspects general indicators sometimes conceal important differences within countries. However, OECD is attempting to develop and apply a set of indicators measuring the pressure of agriculture on the environment (OECD, 1997). Some indicators showing the differences in environmental state between the partner countries are reported below.

1.4.2.1. Land and soil degradation

Data on land and soil degradation are not easily compiled: only some indirect indicators are available. The shifts in percentage of arable land (fig. 1.3) gives a broad indication for the danger of soil erosion as arable land is exposed to greater erosion risks than grassland or permanent crops, while the use of agricultural machinery (fig. 1.4) is an indication for the mechanisation in agriculture and indirectly for soil compaction.

Indicators for soil pollution are not readily available as it can have different causes such as accumulation of heavy metals by use of sludge coming from water purification, contamination through polluted water and so on.

1.4.2.2. Water resources

The main issues concerning water resources are degradation of the quality and in some countries also the quantity of drinking water and preservation of water ecosystems. Agricultural practices influence these issues mainly through nitrate and phosphate surpluses causing eutrophication of surface water and nitrate pollution of ground water, run-off of micro-pollutants and chemical components, bacterial contamination coming from accidental leakage from silos, manure storage or waste water treatments, and salination due to excessive irrigation.

Although there is in general a decrease in the use of fertilisers, the level remains high, particularly in the heavy user group of countries: Belgium, Germany, France and the UK (Table 1.5). Another major source of nutrient losses comes from livestock manure. The figures for the livestock density in total and by grassland area, which are an approximation for this, indicate major differences between countries, although there is also wide variation in density within countries.

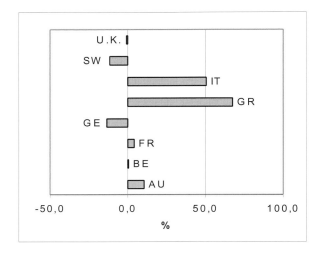

Fig. 1.4. **Changes in the number of tractors per hectare between 1980 and 1995 in the STEWPOL-countries**
Source: OECD (1997)

Pesticide consumption is very high in Belgium, high in Italy, UK and France and to a lesser extent Austria and Germany and low in Sweden and Greece. The use of water for irrigation is a major problem in Southern countries.

1.4.2.3. Impact on landscape and bio-diversity

Agriculture may have two effects which are both negative with regard to landscape:

- In less favoured areas (LFAs), agricultural use may concentrate on the better plots, leaving the marginal areas unmanaged, resulting in a degradation of traditional farmed landscapes.
- In better suited areas, intensification may lead to the reduction of natural landscape elements such as hedges, ponds and woodlands as well as built cultural heritage under the form of traditional buildings, fences, stonewalls, terraces, and so on.

Bio-diversity is also reduced by these two developments through modification of habitats and standardisation of farmland, but the relation between agriculture and bio-diversity is complex and not yet well documented. No universal indicator exists to measure loss of bio-diversity in different countries. As a partial indicator OECD (1997) reports the per cent of threatened and decreasing species in different categories of plants and wildlife.

An important instrument in the protection of bio-diversity is the designation of protected areas or areas with a specific value. Table 1.6 gives an overview of the area in the STEWPOL countries covered by international agreements as listed in the Dobris assessment of 1995 (Stanners and Bourdeau, 1995). Figures may be over-estimated due to overlapping designations but also under-estimated where national designations are not reported. Not all forms of protection have the same importance, but the figures give an impression of the existence of vulnerable zones in the different countries. The comparison between the Dobris figures and the protected areas as registered by OECD (1997) (last column) is interesting, showing the major importance of international designations in Greece and Austria and, to a somewhat smaller extent, in the UK.

Table 1.5
Indicators about the impact of agriculture on the environment in STEWPOL countries in 1996

	Austria	Belgium	France	Germany	Greece	Italy	Sweden	UK
Mineral nitrogen fertiliser use[a] (kg/hectare of agricultural land)	36	113	80	102	37	59	63	78
Livestock density (LU/hectare)[b]	0.8	3.0	0.6	0.9	0.9	0.6	0.6	0.8
Livestock density per grassland hectare (LU/hectare of grass)[b]	1.4	6.6	1.8	2.9	2.9[1]	3.5[2]	4.1	1.8
Pesticides consumption in 1985 (kg/hectare of arable/permanent crop)[c]	3.5	11.3	5.1	3.1[2]	1.9	6.6	1.3	6.8[1]
Irrigated area (percent of arable/permanent crops in 1996)[a]	0.3	0.1	8.4	3.9	34.6	25.2	4.1	1.8
Abstraction of water for irrigation in 1990 (million m^3)[c]	201	n.a.	4.939	n.a.	4.158	32.203[1]	95[1]	181
per cent of total abstraction in 1990 (%)[c]	8.5	n.a.	12.9	n.a.	82.5	57.3[1]	3.2[1]	1.5

[1] 1980; [2] 1990.
Source: [a] FAO internet page, *http://apps.fao.org/cgi-bin/nph-db.pl?subset=agriculture*; [b] individual country statistical services; [c] OECD (1997).

Table 1.6
Total protected areas under international designation (1000 hectares)—1995

Country	Barcelona Convention	Biogenetic Reserves	Biosphere Reserves	Bird Directive	European Diploma	Ramsar Convention	National Parks	Total (in % of total land area)	per cent of protected area[1]
Austria	0	192	28	0	46	103	197	6.8	28.2
Belgium	0	7	0	431	4	8	0	14.8	2.6
France	266	43	647	661	274	423	353	4.9	11.6
Germany	0	1	1,158	842	61	673	700	9.4	26.4
Greece	24	22	9	192	5	104	69	3.4	2.5
Italy	26	34	4	310	77	57	852	4.6	7.1
Sweden	0	1,152	97	0	461	383	632	5.9	4.7
UK	0	8	44	260	217	215	1,393	8.9	19.8

[1] Protected areas as reported by OECD according to IUCN definition of protected area.
Source: Stanners and Bourdeau (1995) pp. 243; OECD (1997).

Table 1.7
Administrative Structure of STEWPOL countries

	Country	Region	Sub-region	Local
Austria	Federal State	9 Länder	District	Community
Belgium	Federal State	3 Regions	10 provinces	589 communes
France	State	26 Regions	100 départements	36,000 communes
Germany	Federal State	16 Länder	District (Kreise)	Communes
Greece	State	13 Regions	51 Nomos	5,000 communes
Italy	State	20 Regions	Provinces	Communes
Sweden	State	23 Counties & Landsting	–	288 Municipalities
UK	State[1]	68 Counties	500 Districts	Parishes

[1] Consisting of four separate territories—England, Wales, Scotland and Northern Ireland.

1.4.3. Institutional framework for agri-environmental policies

An important aspect when discussing the implementation of agri-environmental measures is the institutional policy framework. Two aspects are of importance for this study: the administrative structure of the STEWPOL countries and the actual implementation of Regulation 2078/92 measures. The first aspect is important because it helps to explain differences in implementation of agri-environmental policies while the actual implementation of the Regulation 2078/92 measures indicates the importance of these measures in the different countries.

1.4.3.1. Administrative structure

An important aspect to understand decision making upon and organisation of agri-environmental policies is the administrative structure of the countries. In Table 1.7 the administrative structure of the STEWPOL countries is schematised. Three groups can be distinguished related to their degree of centralisation, ranging from the federal states of Austria, Germany and Belgium, moderately decentralised countries of Italy and the UK and the highly centralised countries including France, Sweden and Greece.

In Table 1.8 the decision and executive power of states and regions in the three state models is presented. This crude classification hides important details, but it explains main differences in options regarding for example, Regulation 2078/92 measures between countries such as Germany and France. In federalised countries like Austria, Belgium and Germany the state only provides a framework at the legislative level and has little or no executive power, while the regions can implement and execute new legislation or regulations. In centralised countries like France, Sweden and to a lesser extent Greece the state has all the legislative and substantial executive power, while the regions have only limited executive functions. The decentralised countries like Italy and UK fall somewhere in between with greater power and influence delegated to the regions.

1.4.3.2. Implementation of Regulation 2078/92

Although EU-Regulation 2078/92 does not cover all CSPs implemented by the countries, the implementation in the STEWPOL countries give a first indication of the importance of agri-environmental schemes in each country. Table 1.9 gives some detail with respect to the budget

Table 1.8
Decision- and executive-power of State and regions in the STEWPOL countries

	Level	Legislative power	Executive power
Federal Countries (Austria, Belgium and Germany)	State	Framework	Low
	Region	High	High
Decentralised Countries (Italy and UK)	State	Framework or decreasing	Low or decreasing
	Region	High or increasing	High or increasing
Centralised Countries (France, Greece and Sweden)	State	High	High or decreasing
	Region	Low	Low or increasing

Regions in the UK are territories

allocated and the uptake of the regulation in the different countries. This shows the major importance of the regulation in Austria (where it was partly used as compensation for the decrease in prices due to accession to the EU), Germany and Sweden, its moderate importance in France, Italy and Sweden and slight importance in Belgium and Greece where, for completely different reasons, the regulation has hardly been implemented. This is also confirmed by figures in Brouwer and Lowe (1998). It shows that Member States have different concerns in terms of both environmental conservation and agricultural support.

1.5. Outline of the book

As indicated in Section 1.2, the chapters in the book follow to a large extent the organisation of the research project. In what follows the content of each chapter is described in somewhat more detail.

In Chapter 2 the results of a classification of 330 agri-environmental measures implemented in the STEWPOL countries are described. The focus is on the economic nature of these programmes analysing their objectives, administrative implementation, the financing and payment vehicle, the degree of competition between environmental and agricultural output, and so on. Through a number of classification tables this chapter explores the economic and institutional rationale behind the policy instruments studied. The inventory data are used to articulate a didactic model of CSPs, adapting the production possibility frontier to an agricultural system producing the joint products of food and fibre (F&F) on the one hand and environmental goods and services (EGSs) on the other hand. The retaliation between production of these two classes of good is examined on the production possibility frontier in terms of their substitution for each other, for the various different policies of the inventory.

In Chapter 3, a broad selection of inventoried CSP measures or instruments is compared using indicators or criteria such as environmental effectiveness, economic efficiency, innovative character, enforceability, mode of operation, impact on property rights, compatibility with a set of policy principles and objectives. Using correspondence analysis, similarities and differences

Table 1.9
Estimated total costs (eligible for co-funding) of the five-year agri-environmental programme under regulation 2078/92 for STEWPOL countries

Country	Austria[1]	Belgium	France	Germany	Greece	Italy	Sweden	UK
Total estimated cost of 2078/92 programme (MECU)	1,006 (2,789)	34	1,905	1,738	17	n.a.	330	378
Total estimated cost per unit of UAA (ECU/ha)	288.5 (800)	25.1	57.2	96.0	4.3	n.a.	98.7	23.1
FEOGA contribution (MECU)	525	18	661	982	11	513	223	170
Per cent of total FEOGA budget	12.2	0.4	15.1	22.5	0.2	11.8	5.1	3.9
Uptake of schemes (1000 ha)	2,998	4	5,893	5,240	na	1,254	1,067	1,202
Proportion of total area (%)	86.0	0.3	16.8	30.3	na	7.3	31.9	6.8

[1]Figure between brackets is an estimate if high expenditure level of first year is maintained over the five years to 1999.
Source: Boisson and Buller (1996) and Deblitz and Plankl (1998).

between instruments are identified as well as their appropriateness in responding to (country) specific situations. They are mapped in a two-dimensional space and the main characteristics of each identified group of policies are discussed and their advantages and disadvantages reviewed.

In Chapter 4 the full costs of countryside stewardship policies are analysed. The cost of agri-environmental measures consists not only of the remuneration paid to farmers, but also of the implementation, monitoring and enforcement costs of the policies, the so called transaction costs. Thirty-seven of the policies described in Chapter 2 were selected for detailed costs analysis. Their full financial costs are estimated and compared. The transaction costs for agri-environmental measures are substantially higher than for price support policies. On the basis of the available information the links between transaction costs and factors such as the type and organisation of schemes and their scale are analysed. The evolution of transaction costs over time is studied and their implications for policy design are discussed focusing particularly on the desirability of attempting to reduce their incidence but also including the way in which they might substitute for compensation payments and the extent to which EU funding might influence their impact.

Next, in Chapter 5, the results of a farmer survey in the eight countries (a total sample of 2000) are discussed. The aim of the survey is to analyse to what extent the uptake of voluntary agri-environmental schemes depends on measure-independent variables, measure-specific elements and country-dependent factors. Besides factors related to farm structure (size, farm type, income level), socio-economic characteristics of the farmers, the remuneration offered by the schemes, the implementation costs for farmers, the effort required to participate in schemes and the environmental attitude of farmers are also used as variables to explain uptake. The results of this survey are valuable as possible predictors of the results of other schemes.

To study the farm economic effects (or the effect on the output of agricultural commodities) of the uptake of agri-environmental measures, a normative model is developed capable of estimating the shifts in land use and effects on the output of agricultural production in a region where these policies are implemented. This model is described in Chapter 6 and applied to eight different European regions. Although in general the implemented measures have an output reducing effect per hectare, the aggregated effect is not certain as the income compensation provided can keep resources in production that would otherwise remain unused or leave the sector. The regions selected for this analysis not only contain very productive regions, but marginal regions as well. The analysis is illustrated for one of the main regions where agri-environmental measures are applied: Oberösterreich. By removing in the model the premiums and requirements of the agri-environmental programme, the effects of the application of the programme can be observed. The results for this and the other regions are analysed, showing that the output effects are indeed limited but that these programmes stabilise incomes and land values in the marginal areas. In that respect they are better targeted than general CAP-policies.

To further explain the low decreasing effect of the application of agri-environmental policies, an equilibrium displacement model is developed in Chapter 7. This type of model is able to take into account both the obvious output reduction effect of the CSP-prescriptions and the output enhancing impact of providing premiums to farmers. As these two effects counterbalance each other, the overall result is a negligible effect on aggregated output, as is empirically proven by application of the model to one of the main programmes in Austria, which is the country with the largest spatial application of Regulation 2078/92 measures.

Agri-environmental support measures are only allowed by the World Trade Organisation under certain conditions. These will probably become even more restrictive in the future and it is therefore important to analyse how far the actual countryside stewardship policies implemented meet these conditions of non-distortion. In Chapter 8 two approaches are followed. One is to calculate the effect of agri-environmental policies on the Producer Support Estimate (PSE), which measures the flow of funds to agricultural support, offering a broad indicator of support through these policies. It thus measures transfers to producers rather than distortions of trade although the latter ultimately result from support. The other approach is to analyse how far selected individual CSP measures meet the conditions of non-distortion used by the OECD such as transparency, targeting, tailoring, evaluation and monitoring. These criteria are applied on a broad selection of inventoried policies.

The book concludes in Chapter 9 with the major findings of the project and some recommendations for further implementation of countryside stewardship policies. Based on some theoretical considerations and the results reported in the preceding chapters, the actual implementation of agri-environmental schemes as well as the possibilities for improving the efficiency of these policies are discussed. Topics such as the desirability of this kind of policy and possibilities for cost minimisation, the creation of market solutions, the targeting of policies, more effective forms of policy organisation are raised. Conclusions as to the long-term contribution of such policies within a policy framework dominated by the next round of World Trade discussions, are drawn.

1.6. Conclusions

Due to technological progress and the use of external inputs the traditional role of farmers in preserving the resource base of rural areas has declined. On the other hand the (social) demand for the environmental and recreational by-products of these stewardship tasks have increased. But as it concerns mainly unmarketable products, the market mechanism does not correct this imbalances between supply and demand. Public intervention is therefore necessary and justified to a certain level and agri-environmental policies are one of the possibilities to increase countryside stewardship by farmers. The purpose of this book is to further investigate the economic impact of such policies and in particular to present empirical evidence of it by analysing attributes and effects of policies applied in the STEWPOL countries.

The results of this work are important as current and prospective policies continue along the road of remunerating farmers for countryside stewardship services. There is also no doubt that in the forthcoming WTO-negotiations these policies will be analysed very carefully. It is therefore hoped by all authors and research teams that this book may contribute to knowledge about the possible impacts of further extending this type of policy.

References

Boisson, J. M. and Buller, H. (1996) The response of member states: France. In: M. C. Whitby, (Ed.), *The European environment and CAP reform. Policies and prospects for conservation*. CAB International, Wallingford.

Brouwer, F. M. and van Berkum, S. (1996) *CAP and Environment in the European Union: Analysis of the effects of CAP on the environment and an assessment of existing environmental conditions in policy*. Wageningen Pers, Wageningen.

Brouwer, F. M. and Lowe, Ph. (1998) *CAP and the rural environment in transition—a panorama of national perspectives*. Wageningen pers, Wageningen.

Colman, D. R. (1994) Ethics and Externalities: Agricultural Stewardship and other Behaviour: Presidential Address, *Journal of Agricultural Economics*, 45(3), 299–311.

Deblitz, C. and Plankl, R. (1998) *EU-wide synopsis of measures according to regulation 2078/92 in the EU*. Federal Agricultural research Centre (FAL), Institute of Farm Economics, Braunschweig, Germany.

Ferro, O., Merlo, M. and Povellato, A. (1995) Valuation and Remuneration of Countryside Stewardship performed by Agriculture and Forestry. In. G. H. Peters and D. D. Hedley (Eds.), *Proceeding of the XXII International Conference of Agricultural Economists, Harare, Zimbabwe*. Dartmouth, London.

Newby, H., Bell, C., Rose, D. and Saunders, P. (1978) *Property, Paternalism and Power*. Hutchinson, London.

OECD (1996) *Economic accounts for agriculture*. Organisation for Economic Co-operation and Development. Paris.

OECD (1997) *OECD environmental data. Compendium 1997*. Organisation for Economic Co-operation and Development. Paris

Stanners, D. and Bourdeau, Ph. (1995) *Europe's environment: the Dobris Assessment*. European Environment Agency, Copenhagen.

G. Van Huylenbroeck and M. Whitby
Countryside Stewardship: Farmers, Policies and Markets
© 1999 Elsevier Science Ltd. All rights reserved

CHAPTER 2

The economic nature of stewardship: complementarity and trade-offs with food and fibre production

Paola Gatto and Maurizio Merlo

Abstract—The paper develops a theoretical framework for the classification of CSPs. The provision of Environmental Recreational Goods and Services (ERGSs) as well as Bads and Disservices (ERBDs) linked to farming and the related environment is analysed according to the theory of public/private goods and the Production Possibility Curve. The definition of reference points and boundaries between positive and negative externalities made up by farming, forestry and the related environment is discussed. It is a critical issue in policy design and implementation and is not yet resolved in Europe. The paper continues with the presentation of the methodology and the results of a survey of 351 CSPs in the eight STEWPOL countries. It is shown that the core of European CSPs is formed of multi-objective and voluntary measures which are also usually temporary and compensated through public funds. The role of long established CSPs, which are included within property rights and very near to ethical values, is also stressed. The paper concludes by showing how current CSPs are affected by unclear definition of objectives as well as ill-understood boundaries between positive and negative externalities. It is, however, acknowledged that current CSPs have initiated a process that in the medium-long term should allow a more comprehensive and effective environmental policy for rural areas.

2.1. The rationale of Countryside Stewardship Policies (CSPs)

2.1.1. What is Countryside Stewardship?

The term Countryside Stewardship (CS) has recently entered, or better reappeared, in the vocabulary of those involved with rural matters. CS generally means a series of operations aimed at land conservation, provision of rural infrastructures, and other positive externalities like recreation and quality landscapes, which are also called Environmental Recreational Goods and Services

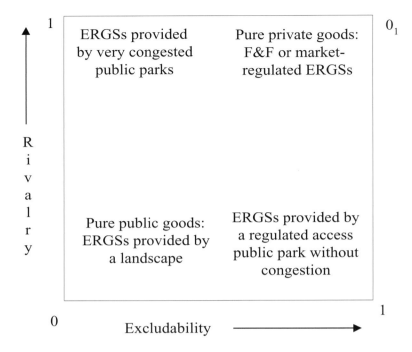

Fig. 2.1. **Classification of ERGSs according to excludability and rivalry.**

(ERGSs). CS is also expected to prevent negative externalities like soil and natural resources degradation/depletion, which are referred to as Environmental Recreational Bads and Dis--services (ERBDs). As Ferro *et al.* (1995) pointed out, the attitude towards CS, and its perception, has certainly changed over time. While in the past the rural population had always carried out stewardship practices with the purpose of conserving the agricultural productive base both for individual owners and for the rural society, CS is now consciously aimed at producing ERGSs and preventing ERBDs. In other words CS is now seen as an entrustment to agriculture and forestry to be an input to the whole rural fabric affecting people's welfare and that of society as a whole.

ERGSs and ERBDs are perceived both as externalities and public goods and/or bads produced by farming/forestry. This double faceted conception is well rooted in economic theory as stressed by Samuelson (1954): 'external effects are basic to the very notion of collective consumption goods'. However, the idea of externalities as public goods is sometimes questioned on the basis of their exhaustibility and rivalry. Only undepletable externalities like those produced by a landscape should be seen as public goods and bads, while exhaustible ones should not (Bonnieux and Desaigues, 1998, p. 31 and Baumol and Oates, 1975, pp. 19–23).

The concept of externality implies policy intervention through internalisation as suggested by Pigou (1920),[1] while the idea of public goods and services opens up the perspective of transformation and development through production and market price as suggested by Coase (1960). Both concepts can be illustrated through the classic diagram of the public/private goods as shown in fig. 2.1. On the lower left hand corner (0) pure public goods and externalities are

---

[1] Internalisation is however questioned by a certain conception of farming, particularly applied to forestry—the so called *Kielwassertheorie* (Dietrich, 1941)—externalities as such should not be rewarded, being a mere function or a 'wake', of the primary production: timber.

located: people cannot be excluded and rivalry is not felt (e.g. a landscape visible from a public road), while on the upper right hand corner (0_1) pure private goods are located: full excludability and full rivalry regulated by the market—e.g. Food and Fibre (F&F) for which consumption is reserved to those paying the market price. Pure public goods and pure private goods, however, represent 'polar cases' as stressed by Samuelson (1955), and the real world situation is much more articulated as argued by various authors trying to criticise or complete Samuelson's model. In particular it has been stressed how excludability can vary according to the property rights, the local customs and the transaction costs of enforcing exclusion. The degree of rivalry is equally variable according to divisibility, rationing and congestion.

The category of mixed club goods, midway between private and public, has been introduced by Buchanan (1967, p. 187) to describe goods which are more efficiently managed and used by a limited group of individuals: so called 'club goods'. 'A club is an organisation which offers shared collective goods exclusively to its members, defraying the cost of the good from members payments' (McGuire, 1987, p. 454). Meanwhile, Tiebout (1956) previously introduced the category of local public goods where exclusion and rivalry can be determined by people's location and local government management—e.g. access to a park and a recreation area. Both local public goods and mixed goods can be located in various sites of fig. 2.1, and they can move around the diagram according to their excludability and rivalry. These cases particularly apply to ERGSs and ERBDs as shown by Buchanan (1967), exemplifying local ERGSs like hunting, then by Randall (1987) and various others environmental economists making reference to ERGSs like those shown in fig. 2.1.

2.1.2. The Production Possibility Curve (PPC) between Food and Fibre (F&F), Environmental Recreational Goods and Services (ERGSs) and Environmental Recreational Bads and Dis-services (ERBDs)

A better understanding of the relationships between F&F, ERGSs and ERBDs as outputs of farming and forestry, and the underlying CS, can be achieved through an hypothetical PPC. The approach outlined in fig. 2.2 emphasises the joint production nature of F&F, ERGSs and ERBDs as proposed by some authors (e.g. Bowes and Krutilla, 1989 pp. 57–58) and Bonnieux and Desaigues (1998, p. 24). It is a useful approach for a descriptive classification of CSPs on the basis of their effects on farming output. Complementarity, indifference or competitive relationships can be singled out. One should, however, be aware of the limitations of this taxonomic approach. First, only one pair of outputs can be considered at a time, while the complex reality of agricultural/forestry activities is such that more than one ERGS/ERBD can be provided at once. Second, this approach is consistent only in a static view of the PPC, while it has no significance when trying to look at relationships between F&F and ERGSs/ERBDs in a changing dynamic context. In other words, the PPC in fig. 2.2 must be seen as a means of classifying CS policies and measures. It does not refer to individual farms and production processes. The dynamic context of an individual farm can only be seen making reference to pairs of outputs: one product and one externality each time.

The PPC of fig. 2.2 describes on the right-hand side traditional farming/forestry resulting from the production processes including various degrees of CS, and on the left-hand side intensive production of F&F without CS. In any case, F&F, ERGSs and/or ERBDs are seen as multiple or joint products. Joint production of F&F and ERGSs on the right-hand side of fig. 2.2 can either be seen as complementary or competitive, according to the segments AA_1, A_1A_2, A_2B_2, B_2B_1 and B_1B. Meanwhile, intensive production of F&F lacks any CS because it requires

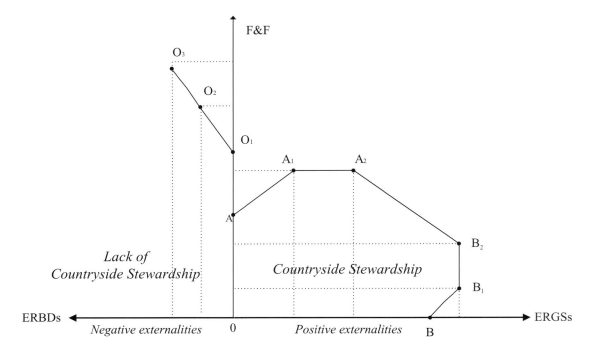

Fig. 2.2. **Hypothetical Production Possibility Curve (PPC) of F&F, ERGSs and ERBDs.**

environmentally damaging operations, with resulting ERBDs. This is shown on the left-hand side of fig. 2.2, in the segments O_1O_2 and O_2O_3. It must be kept in mind that the position of the point 0—therefore of the y-axis—implies a distinction to be made between positive and negative externalities, in other words of boundaries between ERGSs/ERBDs. This is however rather difficult to define, as discussed in the next section.

While CS always remains an input of the production function on the right-hand side of fig. 2.2, it is the economic nature of ERGSs which changes along the PPC. To outline different cases, and contexts, at least five segments can be plotted:

(i) the first production possibility along the positive slope—segment AA_1—is entirely aimed at F&F, which benefits from lower costs, when traditional CS practices such as drainage, hedgerow care, path and road maintenance, crop rotation and so on are carried out. Meanwhile, CS gives rise, almost incidentally, to amenity-related ERGSs such as landscape and footpaths, at negative marginal cost, and this results in maximum complementarity between F&F and ERGSs. Moving from A to A_1 production efficiency increases. It could be interesting to note that the segment AA_1 of the PPC, now almost ignored by textbooks of economics and agricultural economics was well shown in the classic textbooks such as Heady (1952) and De Benedictis and Cosentino (1979). The positive slope was justified, for instance, by complementarity between pasturing of wheat by cattle, which may cause the grain to stool better (Heady, 1952, p. 224); or by crops able to fix N_2O and enhance cereal productivity (De Benedictis and Costentino, 1979, p. 119);

(ii) the second production possibility along the segment A_1A_2 registers amenity-related ERGSs as incidental, or unintended by-products. This is the case for agricultural landscapes or conservation of biodiversity, which involve no additional cost, only environmental care.

Table 2.1
Production Possibilities of ERGSs/ERBDs joint to F&F

PPC segments	Jointness level between F&F and ERGSs	Economic nature of ERGSs	Type of ERGSs produced
AA_1	Maximum complementarity	Inputs and/or intentional products reducing marginal costs of F&F as far as traditional CS is concerned, unintended by-products as far as ERGSs are concerned	Drainage, road maintenance, erosion control, landscape
A_1A_2	Complementarity	Unintended by-product with nil marginal cost for F&F	Landscape quality, biodiversity
A_2B_2	Substitution – trade offs	Products competing with F&F	Hedgerows enhancing landscape and biodiversity, roads, picnic areas, recreation paths
B_2B_1	Complementarity	Main product while F&F becomes incidental by-product	Agritourism
B_1B	Maximum complementarity	Main or sole product, with F&F becoming by-product (and input to the production of ERGSs)	Conservation and wildlife services, parks, recreational areas
PPC segments	Jointness level between F&F and ERBDs	Economic nature of ERBDs	Type of ERBD produced
$O_1O_2O_3$	Maximum complementarity	By-product with social costs	Pollution with environmental degradation/depletion

It should be noted that in modern processes the segments AA_1 and A_1A_2 tend to disappear: hedgerows, for instance, are eliminated, paths reduced, field sizes enlarged and crop rotation abolished. The final result is the overlapping of point A_2 on A on the vertical axis, at a higher level of F&F production (O_1) due to the larger land base, such as drainage systems requiring less soil and cultivation of hedgerow strips. This is the starting point of fully intensive F&F production as described below;

(iii) moving beyond, to the segment A_2B_2, amenity-related ERGSs progressively take on the status of intentional products, involving trade-offs with F&F. This can be the case, for instance, of hedgerow maintenance with all the related benefits, but also the costs. In fact ERGSs take place at the expense of F&F which decrease. The segment A_2B_2, given a certain provision of resources, points out the technically efficient Production Possibility Frontier of what is generally intended as multipurpose farming and forestry;

(iv) ERGSs have a technical limit in B_2, beyond which inefficiency starts: it is not possible to increase their production and/or decrease their marginal costs. F&F represent mere by-products and, in the extreme, incidental products supplied at nil cost. This happens rather frequently in certain mountain areas where crops have been abandoned and land is left neglected for natural re-afforestation or unmanaged pastures. The transition from B_2 to B_1, involves production loss and technical inefficiency;

(v) reducing production and traditional CS practices still further, from point B_1 to total abandonment of F&F at point B, several traditional ERGSs may disappear, as they require a minimum amount of agricultural and forest activity, that is rural maintenance. Pointing out the nature of this segment, Bowes and Krutilla (1989, p. 72), report the case of a wildlife service, which benefits from a certain level of active management, e.g. timber exploitation and farming.

No-CS situations, rather common in modern farming production processes, are depicted on the left-hand side of fig. 2.2. Production Possibilities indicate situations where intensive production of F&F is such that:

(vi) at O_1 no CS practices are applied, therefore ERGSs disappear and production of F&F increases due to the larger land base. The larger land base, making possible higher production, can be due for instance to hedgerow removal or to subterranean drains substituting open ditches. The production possibility O_1 at 0-ERGSs and 0-ERBD is, however, rather theoretical. Usually, the land investment (for example drainage, field size) implies more use of fertilisers and pesticides to reach sufficient returns to pay for them. Consequently the lack of traditional CS often results in production of ERBDs—particularly nutrients and pesticides releases as well a diminished landscape quality. It is a kind of 'vicious circle': large fields decrease landscape quality, disappearance of hedgerows diminishes ecological niches, and, above all, structural investments require the use of chemicals to achieve sufficient returns;

(vii) at O_2 there is a greater use of chemicals to achieve sufficient returns, with pollutant leakages in the soil and in the air, as well as pesticides residuals on the farm products. In other words, the negative effects, i.e. ERBDs, are fully felt;

(viii) at O_3 production of F&F is pushed even further, with an increase in the quantity of F&F produced jointly with ERBDs. It is often a stage where legal boundaries are not operating or, if they are, enforcement is not possible. Note that intensification can also be detrimental to F&F production implying decreasing yields and returns.

2.1.3. Boundaries and reference points between ERGSs and ERBDs

CS practices—whether consciously or unconsciously undertaken—should be looked at in a wider context, and be able to acknowledge the dualistic nature of modern farming: the production of positive and negative externalities, at the same time—ERGSs and ERBDs. Far from the reductive description of fig. 2.2, modern farming/forestry processes involve relationships between a multiplicity of F&F, ERGSs and ERBDs. Positive and negative external effects overlap. The same farm can be in segments AA_1A_2 as far as certain ERGSs are concerned, yet in segments O_1O_2 as far as certain ERBDs are involved. For instance fertilisation and irrigation can improve the landscape quality while diminishing the environmental quality and the biodiversity. This multi-faceted character of farming/forestry is often at the basis of several contradictions affecting present CSPs in Europe. These contradictions are significant for the poor understanding and definition of boundaries between goods and bads, start and end points: in other words benchmarks.

Some authors have tried to define what is positive and what is negative making reference to legal boundaries, or more precisely property rights (Hodge, 1991). This approach, though appealing for its simplicity, seems just a first step. The singling out of the boundaries between ERGSs/ERBDs, as

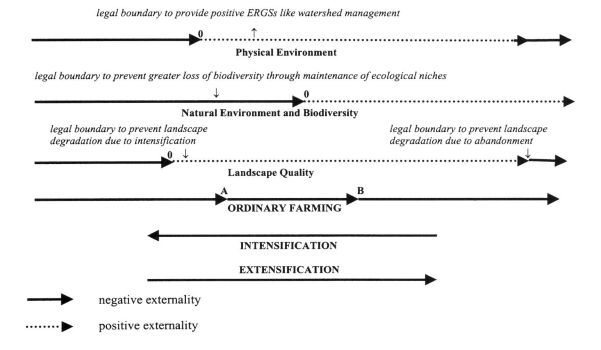

Fig. 2.3. **Boundaries between positive and negative externalities: an attempt to define possible benchmarks and boundaries.**

well as the definition of acceptable reference points, has to be investigated in relation to crops, to the different physical environments and to the various goods and services liked or disliked by the public. An attempt is represented in fig. 2.3, where the three main outcomes of CS—the physical environment, the natural environment and biodiversity and the landscape quality—are shown together with the different possible distinctions between positive and negative externalities, including the legal obligation to provide a certain amount of ERGSs up to the level indicated by [↑] and to prevent ERBDs to the level indicated by [↓].

It should be clear from fig. 2.3, where positive externalities are shown in light shade and negative ones in a darker shade, that intensification implies a diminution of the physical environment, the natural environment and biodiversity and the landscape quality, while extensification acts in the opposite direction. However, it also appears that a certain threshold of extensification can affect negatively the physical environment and the landscape quality and also that ordinary farming as undertaken by the majority of farmers often implies a certain loss of the natural environment and biodiversity. Any definition of boundaries is however biased by subjective judgement, and in any case must refer to contexts where specific reference points should be defined. This operation is far from easy and has never been attempted by CSPs around Europe. Generally speaking, attention has been given to individual aspects, or to groups of homogeneous ERGSs/ERBDs, whereas a more rational and comprehensive view of CSPs is urgently needed.

In order to define farming and forestry 'reference points', or 'politically' acceptable technological packages, the most practical idea, already introduced by some legislation, is to fix reference points according to *good agricultural practices* defined area by area (OECD, 1998, p. 8). At present, codes of good agricultural practices have been issued by governmental agencies with recommendations on a wide variety of farming practices, including fertiliser use, waste disposal, wildlife management and riparian management. Those linked to EU vulnerable areas Regulation 2328/91 should certainly be mentioned though focused on a limited number of objectives.

In summary, this definition process for the identification of benchmarks-boundaries is particularly important in analysing CSPs at a European level. Different perceptions and views should be taken into account: those of experts, those of farmers and those of the public. These perceptions give the basis for defining different boundaries of externalities that cannot be merely linked to legal obligations—which are, generally speaking, emergency thresholds, and cannot therefore be seen as reference objectives. This field of research has been rather neglected until now. Investigation is needed with reference to the different natural, cultural and socio-economic contexts. Legal benchmarks and boundaries certainly remain important: the obligation to provide a certain amount of ERGSs or to prevent the achievement of certain ERBD thresholds. It should be clear, however, at a normative level, that aids and compensation considered by agri-environmental and support measures cannot be linked to legal boundaries alone—there is no ethical and legal point in giving premiums to those whose only merit is respecting the law. Something more has to be done. A much wider range of policy tools should be explored.

2.2. Options and tools for implementing Countryside Stewardship Policies

The analysis of ERGSs or ERBDs provided by farming and forestry, externalities and the related market failures gives rise to two main issues. The first is of a descriptive positive nature: that is understanding, measuring, plotting goods and bads in the rivalry/excludability diagram (fig. 2.1), in the PPC (fig. 2.2) as well as in the positive/negative externalities graphs (fig. 2.3). The second, which is based on the evidence derived by the first, has a political normative nature: that is how the problems of public goods and bads and externalities arising from market failures can be solved.

Three main sets of policy tools have been developed as shown by fig. 2.4: (i) mandatory regulations; (ii) financial/economic internalisation through subsidies and/or taxes; (iii) market creation for public goods/bads and externalities. Meanwhile, complementary measures of information and persuasion, extension, advice and communication are now considered to be essential for implementing all the above sets of CSPs measures. Cross compliance (or eco-conditionality) is also part of the persuasion measures as well as certain voluntary measures with powers of compulsion. In both cases it is a matter of 'strong' persuasion.

2.2.1. Mandatory regulations

CSPs concerning the provision of ERGSs, and the prevention of ERBDs, have been traditionally implemented through legally binding tools, i.e. measures included within constitutions and laws. The framework of property rights, regulations, environmental standards and licences supported by codes of practices and indicators of standards, and planning/programming should be mentioned. Of course societies' changing ethical and cultural values, common understanding

The economic nature of stewardship: complementarity and trade-offs with food and fibre production 29

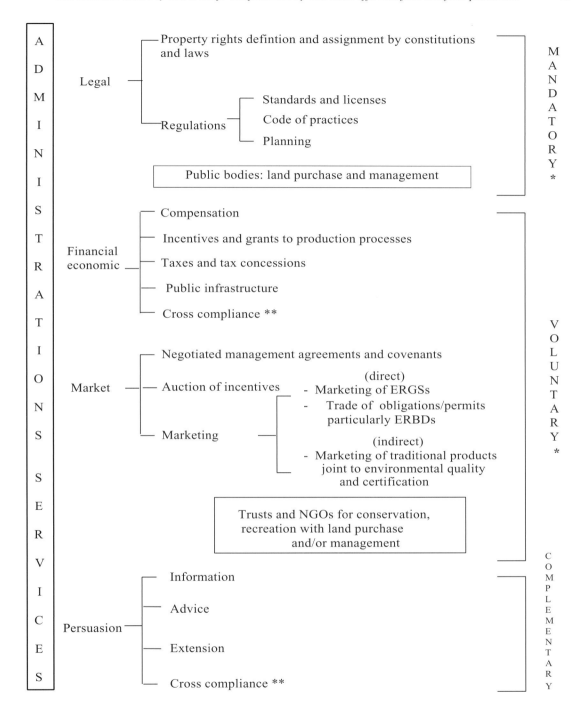

Fig. 2.4. **Policy tools aimed at achieving provision of ERGSs and prevention of ERBDs through CS.**

* In certain countries the difference between mandatory and voluntary measures remains vague: certain mandatory measures should be understood as voluntary with powers of compulsion.
** Cross compliance can be considered as strongly persuasive where the provision of ERGSs is conditional to other financial measures aimed at objectives different from ERGSs—for instance income support for rural development.

and consensus, have always been at the root of legal options. It must also be noted that in certain countries mandatory measures are combined with voluntary ones, and should be understood as voluntary with powers of compulsion.

The adoption of mandatory measures, i.e. the 'stick' approach, which may be softened by some form of compensation, has proved to be an essential part of the policy package aimed at conservation of natural resources and provision of ERGSs as well as prevention of ERBDs. The alternative has been market failure due to 'free riding', fuelling 'tragedy of the commons' types of problems. A negative aspect of mandatory tools, often overlooked, is given by the high administrative transaction costs of policy implementation (Whitby, 1995); and this is particularly relevant when social consensus is not widespread and administration and services are not well established. However, recent evidence shows that as they become established, CSPs show a reduction of transaction costs (see Chapter 4). Another problem has been land abandonment due to lack of profitability, particularly in marginal areas. In these contexts purchase and management of land by public bodies like Public Forest Enterprises has also been much used to guarantee the provision of ERGSs. The experience, in general, is positive, whenever management is supported by effective administration and services. However, neither purchase nor management by public bodies are now considered efficient tools. Costs are generally too high. For instance in Italy forest management costs are estimated around: 200 Euro/ha for states forest; 150 Euro/ha for regional forests; 90 Euro/ha for communal and common property forests, 40 Euro/ha for private forests. Of course it must be mentioned that Italian public forests provide a larger share of ERGSs (Centro Contabilità e Gestione Agraria Forestale e Ambientale, 1998). Semi-public bodies are now preferred such as Consortia, Common Properties, Trusts, and various other Non Governmental Organisations (NGOs).

2.2.2. Financial-economic voluntary tools

More recently, internalisation through subsidies and taxes (the Pigou approach) has been employed as shown by various CSPs, in particular the accompanying measures of the 1992 EU Common Agricultural Policy (CAP) reform and the forthcoming Rural Development Incentives (RDI) of Agenda 2000. This generation of instruments is based on the 'carrot' rather than on the 'stick', in other words positive instruments which are aimed at convincing farmers and landowners to implement certain measures in exchange for various advantages. Quite clearly the financial-economic options follow a Pigouvian rationale (Pigou, 1920) aimed at internalising agriculture and forestry's positive externalities. According to OECD (1996) a 'state pays' approach is applied. The Keynesian side of these instruments, often applied for creating jobs and activating depressed rural economies, is also significant.

Financial-economic options can take various forms according to their economic meaning and administration. Compensation for higher costs and/or lower revenues, due to maintenance of certain land uses and the related ERGSs or prevention of ERBDs, is perhaps the most common tool. Incentives and grants to encourage the production of ERGSs are also applied in various countries. Unlike compensation, the payments are geared to include something more—i.e. a surplus to stimulate participation in a programme. Also tax concessions and taxes, respectively aimed at maintaining traditional farming and forestry and the related ERGSs, or preventing undesirable land uses (ERBDs), have been, and are, much used. Taxes, perhaps the most commonly applied, or at least envisaged, instrument for controlling land use and therefore environmental quality, implement the much advocated, and not yet enough applied, 'Polluter Pays Principle' (PPP).

Also cross-compliance is now a very much advocated CSPs tool, particularly within the EU. Cross-compliance is to a certain extent an indirect financial instrument. The concession of existing payments, such as income support for rural development, is taken as conditional to the adoption of environmentally friendly techniques: i.e. eco-conditionality. In other words, as already stated, cross compliance could be considered a type of strong persuasion supported by various kinds of financial and economic instruments.

2.2.3. Market creation

Marketing through transformation and development of ERGSs is a relatively new CSP tool, envisaged by various countries, and now also supported by the EU rural policies. The approach is, to a certain extent, the one theorised by Coase (1960), and based on the 'beneficiary pays' principle (OECD, 1996). Transformation and development processes can be very different, depending on existing legal and market conditions, but they always have the goal of introducing a certain level of excludability, making consumers pay directly for the ERGSs. The means can vary from institutional or legal to market-led measures.

The transformation towards marketability of ERGS can be complete (to a pure private good) or, partial to a mixed local or club good (fig. 2.1). Management agreements providing payments subject to negotiation between land users and the responsible public authority can be considered the first step (Bishop and Phillips, 1993) toward the market. It is still the 'state pays' approach (OECD, 1996) mitigated, to a certain extent, by the negotiation process. With respect to standard payments they should avoid possible excess payments, resulting in land users' rent (Whitby and Saunders, 1996). At least in theory the negotiation process should approximate compensation or incentives to the marginal cost met by the enterprise, plus the profit necessary to stimulate the agreement. It must be acknowledged, however, that differences with standard payments are far from being well analysed and understood by agricultural administrations.

A compromise solution between the compensation agreed through individual management agreements and the standard payments given to all enterprises applying for them, is given by standard payments differentiated according to sites and specific practices (Ferro *et al.*, 1995). A more extended view of management agreements, requiring contract registration, is given by the so-called covenants, or *servitutes praediorum*, legally attached to the land. From the community's point of view they represent a stronger commitment to the provision of ERGSs or prevention of ERBDs. Auction of incentives is another market-led tool based on competitive bidding directly submitted by landowners wishing to start CS environmentally friendly farming and forestry or to enter an ERGSs production programme. Bids represent the amount of payment for which land owners or users are willing to start the schemes and to accept the restrictions imposed by the programme.

Pure marketing (or commoditisation), according to a Coasian approach (Coase, 1960), foresees specific markets where agriculture and forestry related ERGSs can be sold directly to consumers after transformation into recreational and environmental products. Definition and assignment of property rights is certainly the crucial issue. In order to create a market some institutional changes are needed, in particular assignment of certain rights over the land producing ERGSs (to landlords, the Community, Trusts or other managing bodies) must take place. Recent evidence provided by a hundred European cases of transformation or development of ERGSs into recreational and environmental products shows that institutional changes are important, but are usually not very dramatic (Merlo *et al.*, 1998). On the contrary the key towards marketing is mainly given by the consumption complementarity between market products (recreational and sport facilities, sponsorship, cultural activities, etc.) and the environment where they are produced.

A clear case of consumption complementarity is given by traditional agricultural products, whose image is connected with the quality of the environment and the landscape where they are produced. It is a new CSPs option for which the potential is far from being explored, let alone exploited. This policy tool can be applied whenever direct selling of ERGSs is impossible or 'politically' unwise. Remuneration can be achieved by selling traditional quality products, with the price being influenced by environmentally friendly practices and more generally by the environment in which they are produced. Long established experiences in this field should not be overlooked, e.g. *appellation d'origine controlée* (AOC) of agricultural products (Ferro *et al.*, 1995; OECD, 1996). The now much discussed timber and forestry certification can be conceptually included within these marketing strategies for achieving CS as proposed by Ferro *et al.* (1995). Sustainable management, and CS, could therefore be achieved through voluntary adherence of enterprises to certification schemes which in perspective could create favourable market impacts as a result of differentiation and competitive advantage.

Other full market options are given by agro- and eco-tourism which can be used to achieve CS as they allow remuneration and conservation of unique landscapes. Given the growth of 'green consumerism', the sophistication and willingness to pay of consumers, these CSPs seem to have a future, and therefore an impact on environmental and landscape conservation. Along these perspectives, trusts and NGOs provide conservation and recreation by purchasing land or managing public land. They are now considered powerful tools, and as the real alternative to public purchase and management: to a certain extent the modern 'market' version of ancient common properties traditionally aimed at the provision of public goods for rural village communities.

2.3. The survey of CSPs

2.3.1. Methodology

In order to analyse the economic nature of CSPs at a European level, a survey of CSPs was carried out in 1996 and 1997, using a specific questionnaire. The selection of CSPs was based on criteria derived from the definition of CS, namely:

(i)   relationship with agriculture, and in particular with farmland, therefore affecting directly or indirectly agricultural production processes;
(ii) consideration of one or more of the following objectives: (a) agricultural crops, rural landscapes and related typical elements; (b) natural and semi-natural environments; (c) wildlife and biodiversity; (d) soil conservation and watershed management; (e) recreation and access; (f) reduction of negative impacts of agriculture on the landscape and environment; (g) local quality labels (*appellation d'origine*) with positive landscape impact; (h) afforestation of farmland;
(iii) attention was given to all relevant national CSPs plus a significant sample of regional policies, while local initiatives were considered when particularly meaningful to unique situations;
(iv) representation as far as possible of the variability amongst policy measures, programmes, regional and local situations and typologies in order to reach the maximum possible inclusion of CSPs.

A broad range of information sources have contributed to the survey, in particular legislation, plans and programmes, policy documents, financial expenditures and other provisions, including

Table 2.2
Number of CSPs surveyed and coverage

Country	Number of CSPs surveyed	Inventory level	Sample regions surveyed	% on the total country land area
Belgium	55	National	–	100
Germany	118	Regional	Baden-Wüttemberg, Rheinland-Palatinate, Saxony, Brandeburg, Lower Saxony, Schleswig-Holstein	40
UK	10	Sub-National	England, Wales	62
Italy	71	Regional	Veneto, Emilia Romagna, Lombardia Puglia, Toscana, Abruzzo	40
Sweden	14	National	–	100
Greece	20	Regional	Sterea Ellas Nissi Egaeou, Ipiros Peloponissos Nissi, Makedonia Thraki	45
France	22	National	–	100
Austria	41	National	–	100
Total	**351**			

budgets and year-end accounts. Where written information and official data were not available, 'focused' interviews with stakeholders, officers and professionals have completed the survey. In total, 351 different CSPs from the eight STEWPOL countries project were surveyed.

There was a great difference in the number of CSPs surveyed in each country. Belgium and Austria have compiled a national review, covering almost all of the existing CSPs at various administrative levels, so that all the country was studied; France and Sweden have surveyed the most important 'framework legislation' at a national level, therefore also in this case all the country can be considered included in the survey. On the contrary, the UK has carried out a sub-national review (England and, to some extent, Wales have been surveyed, but Scotland has been omitted), while Italy, Germany and Greece have considered only the most important CSPs at a national level and have subsequently focused on the CSPs in some sample regions.

The variability is due to various causes: administrative structures, centralisation or devolution to regional and local authorities, different approaches to CSPs including the role assigned to agricultural administrations and to environmental and planning authorities outside agriculture. These differences have certainly influenced the results of the survey. In some countries it can be considered as a sample inventory, while in other countries—where devolution of CSPs is more pronounced—the survey provides a representation of regional situations not always significant nation-wide (Table 2.2). For an overview of inventoried CSPs see the Annexe of this book.

The unit of analysis also influences the interpretation of the CSPs survey. The unit could either be a comprehensive policy or individual measures as parts of broader policies. This means that most results and data should be taken only as qualitative and not as quantitative information. A shortcoming of this approach is that all CSPs surveyed, whatever their impact is in terms of participation, receive an equal weighting in the analysis which may introduce bias. In order to overcome misunderstanding, the number of farms and/or hectares has been reported, where possible, to quantify the individual CSPs or CSP-measures. However, in most cases, overlapping of CSP-measures occurred so much that it was not possible to ascertain the total area of all the measures without double counting. It must also be noted that, for about 30% of the CSPs surveyed, number of farms and areas were not available.

2.3.2. Objectives of CSPs

The identification of CSP objectives has proved to be rather difficult, since they are not always clearly stated in legislation or policy documents. Interviews with stakeholders, administrators and officers sometimes highlight 'undeclared' objectives: e.g. income support through the implementation of CSPs.

Initially, all stated objectives were considered without taking into account the predominance of one or other. This analysis has shown that about one third of the CSPs surveyed has one single objective, while half have two or three objectives. This multi-objective nature of CSPs around Europe is interesting but not surprising. The situation could not have been different given that CS operations are generally aimed at creating several ERGSs at one time. Looking at the specific data per country, the situation of the stated objectives of CSPs is more diverse, showing significant differences between countries like Germany, Sweden or France, where the majority of CSPs have 1, 2 or 3 objectives and Greece, where several CSPs have up to 5 objectives at a time. One French CSP has as many as 7 objectives.

The most frequent objectives of European CSPs (Table 2.3) are the reduction of the negative impacts of agriculture (60% of cases), the conservation of wildlife and biodiversity (48%) and the conservation of rural landscapes (37%). The maintenance of a basic environmental quality (in other words prevention of negative impacts, i.e. ERBDs) is the major concern of CSPs all around Europe. Only in Italy and France is this objective overcome by the provision of ERGSs such as the conservation of the rural environment.

Policy measures, therefore, are mainly aimed at situations located on the left-hand side of fig. 2.2, where production (F&F) and negative externality (ERBDs) are concerned. However, the same CSP often includes, at the same time, measures aimed at preventing production of ERBDs and

Table 2.3
Frequency of objectives in the CSPs (*)

Objective	Number of CSPs	% of total CSPs surveyed
1. Landscape conservation	131	37
2. Environment conservation	109	31
3. Wildlife conservation	164	48
4. Soil conservation	74	21
5. Recreation	35	10
6. Reduction of negative impacts	212	60
7. Quality labels	38	11
8. Afforestation of agricultural land	10	3

(*) For more detail see the Annexe of this book

increasing provision of ERGSs, as is the case with Regulation 2078/92. This situation, compared with past stewardship policies mainly aimed at single objectives, allows a more comprehensive approach through which it is easier to detect possible contradictions and enhance synergy between policies.

2.3.3. Rationale and attributes of CSPs

The rationale of CSPs is a very important feature for both classification and policy design and implementation. The main classification scheme is derived from fig. 2.3, where the rationale of the CSPs surveyed can be traced back to three main features, or attributes:

(i) mandatory or voluntary;
(ii) permanent or temporary;
(iii) compensated or uncompensated.

Mandatory CSPs have to be complied with, without any possible exception. By contrast, voluntary measures have to be accepted by the involved parties. In any case, the two terms should be understood in a broad sense bearing in mind that clear boundaries cannot be defined: sometimes the difference between mandatory and voluntary CSPs remains vague, leading to situations where mandatory CSPs should be understood as voluntary with powers of compulsion. The effect of CSPs on the regime of property rights can be total, as in the case of land purchase, or partial when only some land property rights (e.g. land use, application of technical or environmental standards) must be complied with, while others remain with the landowner.

The duration of CSPs can either be permanent or temporary. It is permanent when it cannot be removed: this usually applies to traditional CSPs which are widely accepted by rural communities as ethical values and now part of well-consolidated legislation and local customs. Temporary CSPs last a specified number of years after which they can be renewed or terminated. Schemes which give the farmer or landowner the possibility of backing out of the agreement whenever he decides to do so, should be considered as temporary. Quite a few cases are reported where permanent CSPs are mandatorily imposed affecting land property rights, without any remuneration being granted. Generally, however, to be accepted these obligations must be customary and consolidated within ethical values.

The third substantial feature of CSPs is compensation. Both with mandatory or voluntary CSPs, compensation can be provided to meet increased costs or lower revenues for the farmer or landowner. Compensation for the provision of ERGSs or prevention of ERBDs can be given directly, through financial instruments generally using public funds (the so-called state-pays approach), or indirectly through the market (the beneficiary-pays approach). Payments can be defined within various ranges through a negotiation process: from standard subsidies to graded compensation or incentives according to area, livestock units or yields, to payments negotiated with groups of participants or with individuals in the first step towards a market approach.

Table 2.4 reports the number of CSPs distinguished according the above three attributes. The majority of CSPs are voluntary: 317 out of 351. Thirty-four cases of mandatory measures have been surveyed, mainly concerning traditional long-standing CSPs like the Italian so called 'watershed management bond'. As regards duration, 317 CSPs are temporary, while 34 are permanent. The distribution of permanent and temporary CSPs is different for mandatory and voluntary CSPs: 97% of mandatory CSPs are permanent, while 99% of voluntary CSPs are temporary. Most CSPs involve some form of remuneration (323 out of 351), with a different proportion amongst mandatory and voluntary. While about half of the mandatory CSPs are remunerated, nearly all the voluntary CSPs provide a financial compensation to induce participation. These cases mainly reflect EU Regulation 2078.

Table 2.4
Attributes of the CSPs surveyed

CSP legal basis	Total number of CSPs in the survey
Mandatory CSPs	34
Voluntary CSPs	317
Permanent CSPs	34
Temporary CSPs	317
Uncompensated CSPs	28
Compensated CSPs	323

Table 2.5
Policy tools employed by the CSPs

Financial	Number of CSPs	Market	Number of CSPs	Persuasion	Number of CSPs
Tax concessions	4	Negotiated management agreements	11	Experimental and demonstration farms	25
Credit facilities	4	Negotiated covenants (*servitutes praediorum*)	3	Advice, extension services and training	84
Compensations (total and/or partial)	265	Auction of incentives	–	Information and communication	47
Incentives (including surpluses)	165	Origin labels and certification	15	Cross compliance	5
Levies	4				
Total financial	**442**	**Total market**	**29**	**Total persuasion**	**161**

The predominance of voluntary measures over mandatory ones occurs in all countries. A mixed solution is given by voluntary measures with powers of compulsion, supported by various forms of financial incentives. In all surveyed countries, CS, in contrast with agricultural policies seem to be greatly influenced by the general property rights regime, environmental legislation (e.g. parks and natural resources) and, above all, by physical land use planning by local authorities who regulate building, industry, and other activities including farming.

The financial, market and persuasion measures employed by the CSPs are shown in Table 2.5: 442 cases of financial instruments, 29 market and 161 persuasion instruments have been counted. Amongst financial transactions, 265 cases include compensation according to the spirit of Regulation 2078: payments should make up for higher costs and/or lower revenues to attain stewardship. However, a further 165 cases involve incentives, i.e. compensation plus a surplus to stimulate participation in a certain programme. Obviously the total number of various forms of payments occurring shows that there is some overlapping of these instruments in the CSPs

Table 2.6
CSPs according to administrative levels

Administrative level	Number of CSPs	% of total CSPs surveyed
National/federal	107	30
Subnational	10	3
Regional	220	63
Local	14	4
Total	**351**	**100**

surveyed. In the majority of cases market transactions are given by management agreements where remuneration is still paid by the state, but the approach is mitigated by negotiation, and this means a first step toward the market. On the other hand, origin labels and certification represent the only case in which it is the beneficiary who pays for the ERGSs. Persuasion measures have been applied to support about almost one third of the CSPs. The most significant action being given by extension services and training (84 cases), followed by information and communication (47 cases).

2.3.4. Administration

The survey highlights (Table 2.6) that almost one third of CSPs are managed at a Federal or National level. It is, however, the Regional Authorities that play a key role in CSPs, being responsible for the implementation of the majority of policies (63%). This is not surprising given the local nature of ERGSs and the related CSPs. Furthermore, the survey has shown that a large majority of CSPs is not given by isolated measures but foresee agricultural programmes for implementing CSPs. The rationale of CS applied through comprehensive programmes, seems to go far beyond the well-known Regulation 2078, and clearly shows that CS practices should be—and are—included within wider packages applied at area level. This approach is in line with the required comprehensiveness of CSPs.

Table 2.7 considers the issue of the different authorities concerned with policy administration: implementation, examination, control and monitoring. A remarkable total average number per country of 25 different types of authorities are involved in the different stages of policy implementation, examination and control, for both central and local administrations. These figures, partly justified by the complexity of the problems involved, certainly highlight the role played by bureaucracy and of the related transaction and administrative costs, as often advocated by farmers and landowners. These costs are far from being negligible, although they tend to fall over time (see Chapter 4).

A crucial issue of policy administration is given by the different sources of funding. Table 2.8 underlines the role played by the EU, that is now financing 69% of CSPs surveyed, which generally provides around 50% of the total money involved. The member states are financing 65% of CSPs, although their financial involvement is less than 50%. The Regions, though much involved with policy implementation, play a less visible role in terms of funding. It must be remarked, however, that local Authorities and farmers or landowners have the ultimate responsibility of CS application. The discrepancy between those financing, those implementing and those undertaking CS certainly warrants further analysis. The survey amongst farmers has illustrated this crucial point (see Chapter 5).

Table 2.7
Average number of different authorities involved in the management of CSPs

CSPs procedure	Average number of different authorities
Implementation	
national/regional	9.5
local	8.6
Examination	
national/regional	9.3
local	8.3
Control/monitoring	
national/regional	8.0
local	11.5
Total number of different authorities	**199**
Average number of different authorities per country	**25**

Table 2.8
Extent of funding from EU, member States and other bodies

Share	Number of CSPs					
	EU	State	Regions	Local Authorities	Others	Participants
0–25%	10	118	62	13	1	11
26–50%	162	89	70	3	1	9
51–75%	60	6	5	–	–	1
76–100%	10	14	10	12	4	–
Total	242	227	147	28	6	21
% on total CSPs surveyed	69	65	42	8	2	6

2.3.5. Target areas, uptake and average remuneration

Target areas are usually specified in legislation and various policy documents and programmes. As shown in Table 2.9, the majority of CSPs refer to whole regions or to designated areas; often they also specify a certain land use. Very local issues, on the whole, seem to be often included within more general CSPs, and this can be seen as a positive feature—CSPs are not segmented and flexible application case by case is envisaged.

Problems arise when uptake by farmers and areas must be quantified, since information is sometimes scarce. Table 2.10 reports the uptake resulting from the survey, of course keeping in mind the coverage in terms of the total country area as shown by Table 2.2 and related comments. Uptake is differentiated according to mandatory and voluntary CSPs, number of farms, hectares and funds assigned.

There is a large difference between the number of farms and areas that have taken up CSPs. The application of mandatory measures is outstanding for certain countries: the case is particularly evident for Italy where the so-called 'watershed management bond' is affecting most of

Table 2.9
Target areas of the CSPs surveyed

Target areas	Number of CSPs	% of total CSPs surveyed
Whole country	51	14
Whole Region/Lander	174	50
Mountainous or LFAs areas	10	3
Geographic areas (eg. river banks etc.)	17	5
Local administrative areas (e.g. municipalities)	10	3
Designated areas (e.g. parks, etc)	89	25
Total	**351**	**100**
Specific land uses (e.g. arable, meadows, etc.)	167	48

Table 2.10
Uptake of mandatory and voluntary CSPs in number of farms, hectares and funds assigned

Legal nature	Belgium		Germany		UK*		Italy		Sweden		Greece		France		Austria	
	N	%	N	%	N	%	N	%	N	%	N	%	N	%	N	%
	Number of CSPs surveyed															
Mandatory CSPs	6	11	5	4	2	20	8	11	2	14	2	10	1	5	8	20
Voluntary CSPs	49	89	113	96	8	80	63	89	12	86	18	90	21	95	33	80
	Thousand of farms															
Mandatory CSPs	62		n.a.		3		n.a.		n.a.		n.a.		0.3		n.a.	
Voluntary CSPs	17		585		14		99		n.a.		11		173		815	
	Thousand of hectares															
Mandatory CSPs	849		—		—		9,769		n.a.		n.a.		44		0.1	
Voluntary CSPs	392		—		—		419		1,536		52		7,613		5821	
	Millions of ECU															
Mandatory CSPs	41		3		46		n.a.		n.a.		~0		0.3		1.8	
Voluntary CSPs	11		491		299		163		134		0.06		1,187		864	
	ECU per farm															
Mandatory CSPs	662		n.a.		15,330		n.a.		n.a.		n.a.		1,000		n.a.	
Voluntary CSPs	647		839		21,357		1,646		n.a.		5		6,809		1,060	
	ECU per ha															
Mandatory CSPs	48		—		—		n.a.		n.a.		n.a.		7		18,000	
Voluntary CSPs	28		—		—		389		87		1		156		148	

* The two mandatory UK CSPs should be understood as 'voluntary with powers of compulsion'.
— Total area has not been estimated due to difficulties in quantifying the extent of overlapping of the different CSPs.

mountainous and hilly areas. For other countries, meanwhile, the participation in voluntary measures is remarkable, in particular Austria showing 815,000 cases of participation in voluntary CSPs. Given the total number of farms in the country, this means that on average each farm participates in three CSPs. In other countries the participation rate in the surveyed regions is much lower, for instance in Italy it is 0.1 CSPs per farm. Quite clearly Table 2.10 shows that in certain parts of the EU a greater number of voluntary measures are available, farmers are well informed, schemes are well prepared and managed and consequently farmers are more willing to participate. However, it could also be because the farmers participate as they are asked to comply with something they are already doing. This is the basic criticism to CSPs as shown by a recent UK document on ESAs (House of Commons Committee of Public Accounts, 1998). The information of Table 2.10 seems to be confirmed by the expenditures for CSPs reported by Brouwer and Lowe (1998) and by European Commission's review of the implementation of Regulation 2078/92 (Commission of the European Communities, 1997).

It is rather clear that some countries, and not always the largest, are getting the greater proportion of EU CSPs funds—Austria, Finland, Germany. As stressed by the synthesis report of the Thematic Network 'CAP and the Environment' (Brouwer and Lowe, 1998) in these three countries there is a strong popular and official interest in promoting environmentally friendly farming, but there are other factors also. Germany was a prominent participant in the measures which were precursors to Regulation 2078/92 (the extensification, voluntary set-aside and ESA schemes under Regulation 797/85) and had a range of local schemes that could readily be absorbed and expanded within the new Regulation. Finland and Austria in contrast, devised their national programmes in response to Regulation 2078 and the main reason for the large uptake is that, beside the environmental objectives, they have been seen and promoted as income transfer mechanisms to compensate for farm income losses due to EU accession (Kleinhanss, 1998).

Amongst the other countries that have taken much smaller shares of the agri-environment budget, there are some, such as Belgium, Greece and Portugal, that are newcomers to this policy. For others, notably Denmark and the Netherlands, the relatively low expenditures reflect the fact that the payments available under the Regulation are insufficient to attract the participation of farmers much involved with intensive farming. The low level of expenditures in the UK reflects an approach to the implementation of the Regulation that is oriented to the solution of discrete and specific problems of environment conservation in the farmed countryside through circumscribed and limited measures, rather than general extensification schemes (Brouwer and Lowe, 1998), while the high level of payment per farm reflects the larger size of the farms.

2.3.6. Commodities and markets potentially affected by CSPs

The survey shows a certain distribution of commodities potentially affected by CSPs, with a predominance of cereals, root crops and grassland (Table 2.11). Less attention seems to be paid to old olive groves and traditional vineyards, paramount features of Mediterranean landscapes producing so many positive externalities. However, this can be attributed to the limited extension of these crops in the area surveyed—i.e. they can be found only in Italy, Greece and the South of France. In addition, CSPs are aimed, above all, at preventing negative externalities, as shown by the objectives of CSPs listed in Table 2.3, rather than promoting positive ones, like those linked to quality landscapes.

Table 2.12 shows that a remarkable 39% of ERGSs resulting from CS can be marketable, however only 12% are already the subject of some sort of marketing, meaning that CS is deliberately aimed at producing ERGSs, showing profit maximisation behaviour. This option, the most efficient and less expensive for the public budgets, is very promising, and already studied by other EU research projects (Merlo *et al.*, 1998). The limits are given by the difficult marketability

Table 2.11
Agricultural markets potentially affected by the CSPs

Agricultural commodity	Number of CSPs	% of CSPs surveyed
All agricultural commodities	84	24
Cereals	83	24
Root crops	60	17
Fruit	31	11
Fresh vegetables	24	7
Vineyards	27	8
Olive groves	11	3
Seeds	5	1
Flowers	5	1
Other crop products (mainly grassland)	58	17
Meat	84	24
Milk	59	17

Table 2.12
Marketability of ERGSs resulting from CS

Nature of ERGSs	Number of CSPs	% of total CSPs surveyed
Already marketed	41	12
Potentially marketable	96	27
Not marketable	205	58
Answer not possible	9	3
Total	351	100

of the majority of ERGSs resulting from CS practices—58% of cases as shown by Table 2.12. Clearly, whenever ERGSs involve option—if not existence and non-use—values, the market choice becomes more remote.

2.3.7. Farm practices affected by CSPs: trade-offs between F&F, ERGSs and ERBDs

Table 2.13 shows that CS is mainly related to a change in agricultural practices with 259 cases out of 351. This means that 74% of CSP involves the usual daily operations undertaken by farmers. The preservation of agricultural landscapes is also considered very important: farmers are affected by CSPs when they adopt certain crop types and land uses. Measures aimed at the natural environment or at individual elements of landscape seem to be less considered. Recreation and access only accounted for 11% of the CSPs, although this proportion is likely to increase in the future.

The relationships between ERGSs, ERBDs and the production of F&F are a crucial issue. Table 2.14 shows that almost all of the segments which make up the PPC of fig. 2.2 are, to a certain extent, targeted by CSPs. These relationships can be considered in terms of complementarity, competition

Table 2.13
Technical aspects of stewardship measures

Technical aspects	Number of CSPs	% of total CSPs surveyed
Agricultural landscapes	204	58
Natural environments	87	25
Elements of landscapes/environment	89	25
Recreation and access	40	11
Agricultural practices	259	74

and substitution. Given, however, the extremely relevant role played by CSPs aimed at reducing negative impacts, as shown by CSP objectives (Table 2.3), it is rather clear that much effort has been aimed to prevent ERBDs. In other words, CSPs mainly seek re-introduction of CS practices neglected by modern intensive farming.

Figure 2.5 offers a graphic representation of the data presented in Table 2.14 with reference to the PPC of fig. 2.2, while adding a quantitative evaluation of each CSP in terms of area covered and number of CSPs involved. The positions of the CSPs, as expressed in the survey, have been located in the different sites of the PPC and represented by segments whose length is proportional to the number of CSPs in fig. 2.5 (a) and to the total area covered in fig. 2.5 (b). This presentation provides some quantitative understanding of CSPs. Figure 2.5 (a) shows that the number of CSPs finds a rather balanced representation on the different sites. The inventory has been able to capture a large variety of possible measures, reflecting the proportion of the hypothetical PPC already shown by fig. 2.2. This balance could also reflect the choice of sample CSPs selected in the survey.

Figure 2.5(b) shows that—in terms of area covered—the most frequent relationship between F&F and ERGSs/ERBDs is non-competitive. Segment AA_1, with 48% of the total surveyed area, is the site where the most traditional CSPs are located, e.g. those aiming at soil conservation, maintenance of the rural fabric, with amenity-related ERGSs as unintentional by-products. This result, though remarkable, is not surprising as it is a sign of continuity between past and present CS practices. It means farmers are paid for something they have always done and could do also today without being paid. However one might doubt that these practices like maintenance of grass will continue in future without support, as these practices are under high pressure and could shift from segment AA_1 to the left hand of fig. 2.5 (segment O_1O_3), producing negative externalities. Therefore, promoting traditional styles of farming rather than having to prevent negative externalities of modern practices can be justified. In fact the area devoted to the prevention of negative externalities is remarkable as well, as already in the case of fig. 2.5(a), although in a lower proportion. The smaller areas of segment A_2B_2 where CSP measures signify real trade-offs between F&F and ERGSs reflect the current situation where real multipurpose agriculture and forestry are much discussed and little applied. There are also very few situations where the ERGSs are the main or only products with F&F becoming by-products.

2.4. Conclusions

The main outcomes of the theoretical analysis and the survey of CSPs allow several conclusions to be drawn:
(i) CS should be considered as an input of a multi-objective production function, where ERGSs together with F&F are the products. The lack of CS, except for the few cases of real wilderness, can generate ERBDs;

(a) *length of segments estimated as a percentage of the total number of CSPs surveyed* as reported in Table 2.14 (leaving out the cases of 'answer not possible')

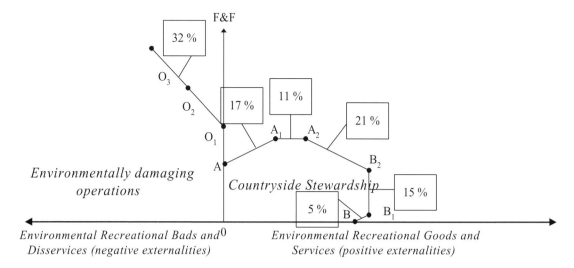

(b) *length of segments estimated as a percentage of the total surveyed area.* The total area surveyed—which the percentages reported in the PPC refer to—is around 160 million hectares. However, the data are heavily biased as for almost 30% of CSPs the information on the land base was not available, while for the remaining 70% with available information, double counting could have occurred.

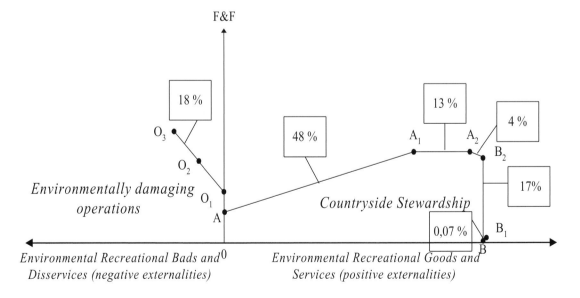

Fig. 2.5. Hypothetical location of Countryside Stewardship Policies (CSPs) on the Production Possibility Curve (PPC)*

* Slopes on the left-hand side have been arbitrarily set at 45°, while slopes on the right hand side are bounded to the vertical and horizontal axes.

Table 2.14
Estimated relationship between ERGSs, ERBDs and F&F (*)

Relationship between ERGSs and F&F	Number of CSPs	% of total CSPs surveyed
Complementarity between F&F and ERGSs		
Segment AA_1	55	16
Segment A_1A_2	36	10
Competition between F&F and ERGSs		
Segment A_2B_2	69	20
Complementarity between ERGSs and F&F		
Segment A_2B_1	50	14
Segment B_1B	15	4
Relationship between ERBDs and F&F		
Complementarity between F&F and ERBDs		
Segment $O_1O_2O_3$	104	30
Answer not possible	22	6
Total	**351**	**100**

(*) For more detail see the Annexe of this book

(ii) CSPs are generally multipurpose policies including several objectives simultaneously: primary and secondary, stated and un-stated, short and long term;
(iii) the need for CS aimed at reducing negative impacts, i.e. ERBDs, is the rationale behind many CSPs;
(iv) modern CSPs are mainly based on voluntary, temporary and compensated measures. Compensation by the state is prevalent;
(v) market measures are, however, emerging, based on the 'beneficiary pays' approach: some 40% ERGSs resulting from CSPs are marketable, opening up new perspectives for beneficiary payments;
(vi) the role of long established CSPs, included within property rights and land use physical planning, very near to ethical values, though outside the main field of agricultural policies, must not be neglected as they represent the institutional framework of any CSP;
(vii) CSPs are mainly managed at regional or local levels, while financial sources come from a higher level; in particular the EU and member states. Implementation is made through programmes, i.e. packages of measures responding to the multipurpose nature of CSPs;
(viii) all agricultural commodities are affected by CSPs. In terms of policy cases, the so called 'grand cultures' (major arable crops), and animal production, are prevalent;
(ix) relationships between ERGSs and F&F are at the heart of CS but they seem far from well defined, jeopardising the development of CSPs.

Unclear definition of CSPs objectives as revealed by the survey, represent the main shortcoming of agricultural and, indeed, environmental policies for rural areas. Difficult and/or ill-defined boundaries of ERGSs/ERBDs, and the underlying potential contrasts, are now a major issue to be solved in terms of definition of good agricultural practices. The process is made difficult by the emerging competition among different ERGSs and ERBDs which represent the future challenge for CSPs. It must be acknowledged, however, that progress has been made in terms of policy comprehensiveness as shown by the survey. The relative scarce results of the agri-environmental measures undertaken in the '90s, due to the novelty of these policies and their long pay-off period, should therefore not discourage further consistent endeavours towards better and more effective CSPs.

References

Baumol, W. J. and Oates, W. E. (1975) *The theory of environmental economics*. Prentice Hall, Englewood Cliffs N.J.
Bishop, K. D. and Philips, A. P. (1993) Seven steps to market: the development of the Market-led Approach to Countryside Conservation and Recreation, *Journal of Rural Studies*, 9(4), 315–318.
Bonnieux, F. and Desaigues, B. (1998) *Economie et politiques de l'environment*. Précis Dalloz, Paris.
Bowes, M. D. and Krutilla, J. V. (1989) *Multiple Use Management: the Economics of Public Forestlands*. Resources for the Future, Washington D.C.
Brouwer, F. and Lowe, P. (Eds.) (1998) *CAP and the Rural Environment in Transition—A panorama of national perspectives*. Wageningen Pers, Wageningen.
Buchanan, J. M. (1967) *The Demand and Supply of Public Goods*. Rand McNally & Company, Skokie, Illinois, U.S.A.
Centro Contabilità e Gestione Agraria, Forestale e Ambientale (1998) unpublished data. University of Padua.
Coase, R. (1960) The problem of social cost, *Journal of Law and Economics*, 3, 144–171.
Commission of European Communities COM (1997) Report from the Commission to the Council and the European Parliament on the application of Council Regulation (EEC) No. 2078/92 on agricultural production methods compatible with the requirements of the protection of the environment and the maintenance of the countryside, Brussels 04.12.1997 COM (97) 620 final.
De Benedictis, M. and Cosentino, V. (1979) *Economia dell'azienda agraria*. Bologna, Il Mulino.
Ferro, O., Merlo, M. and Povellato, A. (1995) Valuation and Remuneration of Countryside Stewardship performed by Agricultural and Forestry. In: G. H. Peters and D. D. Hedley (Eds.), *Proceedings of the XXII International Conference of Agricultural Economists, Harare, Zimbabwe*. Dartmouth, London, pp. 415–435.
Heady, E. O. (1952) *Economics of Agricultural Production and Resource Use*. Prentice Hall, Englewood Cliffs, N.J.
Hodge, I. (1991) The Provision of Public Goods in the Countryside: How Should It Be Arranged? In: N. Hanley (Ed.), *Farming and the Countryside*, CAB International, Wallingford.
House of Commons Committee of Public Accounts (1998) *Protecting Environmentally Sensitive Areas*. 39th Report, 3rd June, Stationary Office, London.
Kleinhanss, W. (1998) New Member States: Austria. In: F. Brouwer and P. Lowe (Eds), *CAP and the Rural Environment in Transition—A panorama of national perspectives*. Wageningen Pers, Wageningen.
McGuire, M. (1987) Public Goods. In: J. Eatwell, M. Milgate and P. Newman (Eds), *The New Palgrave: A Dictionary of Economics*. MacMillan, London.
Merlo, M, Milocco, E. and Virgilietti, P. (1998) In: *Transformation/development of ERGSs provided by forestry into RES-Products*. Final Report of FAIR CT95-0743, EU Research Project on 'Niche market for recreational environmental services'.
OECD (1996) *Amenities for rural development*. OECD, Paris.
OECD (1998) *Reference levels for agri-environmental policy measures: distinguishing between beneficial and harmful environmental effects of agriculture*. Joint Working Party of the Committee for Agriculture and the Environment Policy Committee. COM/AGR/CA/ENV/EPOC(98)54, OECD, Paris.
Pigou, A. C. (1920) *The Economics of Welfare*. MacMillan, London. Italian edition: Economia del Benessere, Torino, UTET (1948).
Randall, A. (1987) *Resource Economics—An Economic Approach to Natural Resources and Environmental Policy*. John Wiley and Sons, New York.
Samuelson, P. (1954) The pure theory of public expenditure, *Review of Economics and Statistics*, 36, 387–389.
Samuelson, P. (1955) A Diagrammatic Exposition of a theory of public expenditure, *Review of economics and statistics*, 37, 350–356.

Tiebout, C. M. (1956) A Pure Theory of Local Expenditures, *Journal of Political Economy*, October n. 5, pp. 416–424.

Whitby, M. (1995) Transactions Costs and Property Rights: The Omitted Variables? In: L. M. Albisu and C. Romero (Eds), *Environmental and Land Issues—An Economic Perspective*. Wissenschaftsverlag Vauk, Kiel.

Whitby, M. and Saunders, C. (1996) Estimating Conservation Goods in Britain, *Land Economics*, 72(3), 313–325.

G. Van Huylenbroeck and M. Whitby
Countryside Stewardship: Farmers, Policies and Markets
© 1999 Elsevier Science Ltd. All rights reserved

Chapter 3

Policy indicators and a typology of instruments

François Bonnieux and Pierre Dupraz

Abstract—CSPs are implemented through a variety of policy instruments which target the achievement of both environmental effectiveness and economic efficiency. Many other factors also come into play such as distributive and trade effects. Criteria for evaluating CSPs are classified into five categories which refer to efficiency, sustainability, enforceability, political acceptability and equity. A series of 70 criteria have been precisely defined and filled for a selection of 143 CSPs distributed among the eight countries of interest. The background information is provided by the prescriptions and requirements or codes of agricultural practice of the various policies. Some criteria are quantitative and some are purely qualitative so to deal with this information multiple correspondence analysis has been used. The basic philosophy is to reduce dimensions in order to reveal relationships among the CSPs, among the criteria and between the CSPs and the criteria. It is found that six factors give an adequate approximation to the data set. These factors are easily interpreted and are used to classify CSPs according to seven groups. Three groups are defined according to the constraints which are put on the use of fertilisers and pesticides, whereas three others discriminate between policies according to the type of goods which are promoted (e.g. local public goods, organic food, labelled products). The definition of the seventh group is more ambiguous and the profile of the CSPs concerned is close to the average policy profile. The distribution of CSPs according to countries and groups shows a country bias which mainly results from the way in which Regulation 2078/92 is implemented in the eight STEWPOL countries. But it is also partly due to the way in which the sample of policies has been selected.

3.1. Introduction

This chapter emphasises the analysis of sectoral issues and focuses linkages between environmental concerns and agricultural policy. The most challenging problem that must be dealt with is determining the most appropriate way of measuring changes in environmental quality and natural reserves due to countryside stewardship policies. In a long run perspective this could be viewed as being a preliminary step to achieving green national accounts (e.g. Adger and Whitby, 1991 or CBO, 1994). The following step, which is out of the present scope, would consist of pricing the non-market services of environmental assets.

Environmental indicators are often defined in a very simple way: 'an environmental indicator is a figure which gives a simple indication of the status of a specific condition, in the environment, and any changes in such' (CBS, 1992). The nitrate concentration in drinking water gives an example of such indicators. Some indicators are associated with threshold values, e.g. the maximum of 50 mg of nitrate per litre recommended by the World Health Organisation. But the important issue is not the threshold value per se but its policy implications, since it requires governments to monitor progress in meeting the conditions of international agreements. So there is a need for figures which give a comprehensive picture of agri-environmental linkages and which are policy-relevant.

A considerable body of conceptual literature has clarified links between environmental impacts, farming practices, and agricultural and environmental policies (Abler and Shortle, 1990; Hanley, 1991; Just and Antle, 1990; Just and Bockstael, 1991; Parris, 1999; Vail *et al.*, 1994). Relations between environmental effects of a scheme can be of a conflicting or of a consistent type leading to difficult problems to deal with. Thus the approach followed by international organisations and public agencies may obscure the social and political forces at work (Moxey *et al.*, 1998). Nevertheless more quantifying work to assist policy makers is needed. Following various authors, indicators are defined as a composite set of attributes (or measures) which embody a particular aspect of agriculture. It is the combination of these attributes which give the indicator. Therefore indicators range from raw data to more elaborate synthetic variables. As indicators are redefined and validated they enhance the level of knowledge around which a broad consensus can be formed.

Four important issues have to be considered: policy relevance, theoretical soundness, measurability, and aggregation relevance. A proposed indicator must serve the needs of decision-makers and its use should lead to better decisions than would have been possible without it. From this perspective, policy relevance refers to the ability to show a clear linkage between agriculture and environmental impacts. This implies the definition of threshold and target values in order to easily interpret trends over time as well as spatial differences.

Theoretical soundness requires that indicators are theoretically well founded. To be valued, they must be based on a consistent theory and derived from an accepted methodology. This implies they result from a comprehensive debate unambiguously related to a strong theoretical background, forecasting and information system. This should improve communication with scientists, landholders and the general public. With respect to the public a reservation must be made. Several case studies (e.g. Bouwes and Schneider, 1979) show significant differences between environmental perception and scientific indicators of environmental quality. Thus communicating to the general public requires a long run effort in terms of extension and education. This cannot only rely on the publication of quantitative indicators.

As regards the measurability issue, indicators must satisfy valid sampling, statistical and consistency methodologies. Attributes from which they are derived, are mainly selected from existing statistical data sets. In this context it is crucial to rely on accessible data with low measurement errors and to select data sets with historical data suitable for trend identification.

Indicators must be applicable at national level and capable of regional differentiation as well as international comparisons. Besides there is a challenge in addressing linkages at the farm level. At the moment there is a limited experience with farm-level monitoring and reporting of environmental conditions. In addition most agricultural data are collected at the level of administrative areas which differ from agro-ecological zones. Therefore the appropriate level of aggregation for each attribute has to be carefully considered.

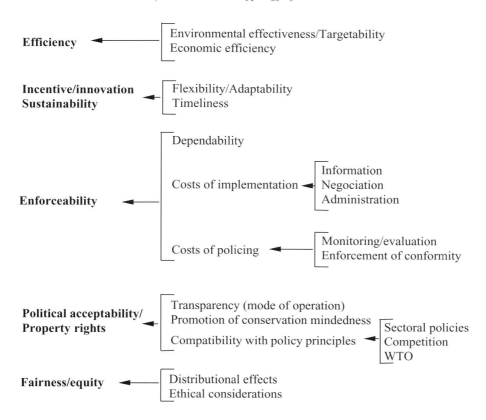

Fig. 3.1. **Criteria for evaluating CSPs.**

This chapter is organised into four main sections and a concluding one. The first two emphasise theoretical issues with a discussion of the relevant indicators to categorise CSPs and a presentation of the methodology. The other sections are devoted to empirical results and policy implications of the proposed typology of the CSPs.

3.2. Definition and selection of criteria

3.2.1. General guidelines

CSPs are implemented through a variety of policy instruments. None of them is a panacea. Some instruments are better adapted to specific environmental problems and economic circumstances. It would be inappropriate, for instance, to impose charges on hazardous substances to protect a water catchment area, as a strict control or ban would be more effective. In all cases one must reach a delicate balance to ensure the achievement of both environmental effectiveness and economic efficiency. Many other factors come into play, such as distributive and trade effects. The choice of instruments should be based on a number of economic, social and political criteria which are classified into five categories (fig. 3.1). Data were obtained from each country for a number of CSP instruments. Parts of this data may embody expert judgements about aspects of the policies which may not be replicable.

Efficiency is a widely used criterion for economic performance. Under ideal conditions a competitive equilibrium is Pareto-efficient. The conditions which gear the farm sector are far from this ideal framework. Indeed there are a number of market failures resulting from price support as well as a series of externalities. Therefore it is not relevant to adopt a first best perspective. In a classical paper, Lipsey and Lancaster (1956) stated that 'it is not true that a situation in which more, but not all, of the optimum conditions are fulfilled is necessarily, or is even likely, to be superior to a situation in which fewer are fulfilled'. This second best theorem has negative implications for welfare economics because intuitive ideas expressed by the Pareto conditions are not theoretically justified. Thus, it is appropriate to consider efficiency from two separate standpoints, each of which suggests different criteria. (i) Environmental effectiveness: to what extent will CSPs achieve environmental objectives? (ii) Economic efficiency: will CSPs achieve goals at minimum costs to society? With respect to the environmental effectiveness perspective the following point has to be considered: are environmental objectives precisely specified? It is a crucial issue because there are significant differences between policies.

Incentive refers to a dynamic aspect which is essential for the future, since the objective is to value policies in terms of sustainability. Do instruments give a continuous incentive to provide environmental goods at a minimum cost? Flexibility (or adaptability) is a key concept. Do the programmes offer incentives to farmers to select the best available technology to produce amenities? Does the policy tell the farmer exactly what to do, or does it offer the broadest scope of choices consistent with environmental objectives? Timeliness refers to a different concept. It focuses on the speed with which a specific CSP can be mobilised to achieve an objective. A division between two categories of policies is expected to be instrumental. The first category refers to policies based on the implementation and enforcement of a new institutional framework, they can establish a market (labelling provides an illuminating example from this point of view). The second one emphasises transitory measures which are implemented to facilitate the adoption of different technology, such as organic farming. Moreover, policies which imply a change in technology are expected to induce a permanent shift in favour of the provision of environmental goods.

Enforceability includes several points. Dependability refers to the reliability of a specific CSP in achieving its objectives. Is its outcome almost sure or does it depend on a number of uncertain factors? Costs of implementation and costs of policing include all components of transaction costs.

Political acceptability depends on many factors such as cost, simplicity, transparency, public participation and consistency with other policies. Therefore this quite heterogeneous category covers three different issues which are linked with political acceptability.

(i) Transparency (mode of operation): policies which are clearly defined in terms of objectives, financial arrangements and control are likely to be more acceptable both to farmers and to taxpayers as mechanisms for the provision of environmental goods.
(ii) Promotion of conservation mindedness: is a specific CSP likely to promote positive attitudes to conservation among farmers and the general public?
(iii) Compatibility with policy principles: CSPs operate within the context of CAP and of many international agreements.

They must be properly integrated into agricultural policy, e.g. control of polluting inputs imply the removal of related subsidies. They must also comply with trade rules within the EU and WTO context.

The latter received special attention in our research. Environmental policy instruments may have socially regressive impacts by increasing farming costs. Otherwise, they may negatively affect consumers' welfare as well as that of taxpayers' through the price mechanism. Does a specific programme divide its financial burden among farmers, consumers and taxpayers fairly? To what extent are regressive income effects compensated? These issues lead to consideration of the compatibility of CSPs with the polluter-pays-principle and the beneficiary-pays-principle. Otherwise, these policies also generate substantial bureaucratic employment which is probably also regressive (Whitby, 1996).

3.2.2. Overview of criteria

All the criteria which are proposed are derived from these general guidelines. They do not rely on value judgements, as they are based on the explicit targets of the CSPs which are examined. It must be emphasised that these targets are generally intermediate objectives and are often loosely linked with ultimate objectives. For instance, several policies aim at reducing fertilisation, which is an intermediate objective as the ultimate one is to decrease nitrate contamination of water bodies. Here the background information is provided by the prescriptions and requirements or codes of agricultural practices of the various policies.

3.2.2.1. Environmental effectiveness

To stimulate the provision of environmental benefits, two options are available, implementing a policy with restrictive prescription to a small area, or a non-restrictive one to an extended area. Real policies result from a trade-off between these two opposite standpoints. The provision of environmental goods, involved by a specific CSP, will depend on two categories of factors. The first one refers to management prescriptions (e.g. reduction of inputs) whereas the second one is related to the scope of the policy (e.g. designation and number of hectares enrolled). Before defining the criteria it must be emphasised that a good indicator discriminates between policies and therefore, does not have a flat profile.

Water protection is a very crucial issue for the future development of the farm sector. Four qualitative criteria are defined in order to appreciate the objectives of the CSPs relative to water protection. The first one addresses the quantity dimension of the resource. The others focus on negative externalities by considering prescriptions which ban or limit fertilisation and pesticides as well as heavy metals or target the control of erosion.

A series of qualitative criteria is defined to assess the expected effects on human health, bio-diversity (fauna and flora), and landscape (aesthetic, cultural heritage and fire protection). The protection of grassland and rough grazing depends on the flocks grazing it. The density of livestock is a synthetic variable which takes this point into account. It is used to define a specific criterion. The impact of policies on animal welfare is an issue in itself. It is captured with a very simple indicator.

A simple indicator has been used to discriminate between policies according to the level at which they are implemented. This indicator is expected to be significant for evaluating environmental effectiveness since eligible area depends upon administrative level. This indicator emphasises the level at which the policy is managed. There is a link with the issue of targetability. In order to estimate the scope of each CSP, three quantitative criteria are considered: the number of farms enrolled, total area covered and the number of livestock units which are concerned. Moreover, compensation and total costs (including administrative costs where available) are also reported.

3.2.2.2. Economic efficiency

Three different issues are considered under this heading. First of all, targetability is assessed with criteria which indicate the category of zoning (geographical, ecological and agronomic) and a criterion which discriminates between policies which apply to the whole farm and policies which do not.

Then the nature of the output is described with a series of indicators. Three indicators emphasise non-environmental outputs provided by CSPs. The first criterion focuses on the marketability of the output provided by the CSPs. The promotion of quality products as well as green tourism concerns a significant number of CSPs and therefore two criteria are defined to categorise them. Moreover two indicators are considered to describe the underlying technology. The first one refers to labour intensity whereas the second one describes the relationship between environmental and agricultural outputs. The former indicates whether the CSP is expected to lead to an increase in the labour force or not. The latter indicates whether the farm commodity and the environmental goods and services are substitutes or complements

In the last point, quantitative variables expressing the cost per farm and per hectare of the selected policies are defined. In order to overcome the difficulties stemming from the lack of information concerning transaction costs a qualitative indicator has been defined. It refers to the share of public administrative costs compared with exchequer costs (total payments received by farmers plus administrative costs) and three classes are defined (share under 10%, between 10% and 50%, and share over 50%).

3.2.2.3. Sustainability

Two main issues, incentiveness and enforceability are successively considered. With respect to the former four criteria have been considered. The first one refers to the length of the contract in number of years and the second one equals the number of farms enrolled each year. The number annually enrolled is indeed a basic variable to appreciate the speed at which policies are implemented. Policies leading to the adoption of a new technology for providing environmental goods and services are distinguished by a specific criterion. Here the focus is on CSPs which promote extensification and lead to a better use of the natural basis of agriculture. Finally, a specific indicator focuses on possible irreversibility of changes involved by CSPs and therefore emphasises one aspect of sustainability.

Four simple dummy variables are defined to deal with enforceability. They describe the way in which CSPs are implemented and controlled. Ease and reliability of the control are especially considered.

3.2.2.4. Political acceptability

This covers the following issues: mode of operation, property rights, compatibility with policy principles and equity. The mode of operation refers to different points which are taken into account with four criteria: transparency, voluntary participation, individual agreement and education. The last identifies policies which are expected to promote conservation mindedness.

A possible restriction put on property rights is considered as the process used to compensate farmers (purchase or rent of property rights) or not (command and control or use of a charge). It must be pointed out that there are some policies with no property right limitation but involving some compensation. This is, for example, the case of the maintenance of hedgerows. A specific criterion describes the type of instrument (market instruments, regulation, economic, financial instruments and persuasion).

The compatibility of CSPs with international agreements is a crucial issue for the future WTO round. A series of indicators are introduced in order to distinguish between a positive effect on the aggregate supply, a negative one and no effect. This criterion is defined for various commodities of interest: cereals, roots, fruit, wine, olive oil, meat and milk. To deal with decoupling, two simple indicators are defined. They address the issue of over-compensation for the restrictions put on farming activities and the relative share of compensation paid in total sales. Simple indicators are also defined to discriminate CSPs which target the correction of a negative externality involved by past policy, which favour extensive farming or the conversion from intensive to extensive farming. The polluter-pays principle (PPP) is one of the basic principle of the EU environmental policy. It is therefore crucial to analyse the consistency of CSPs with the PPP. This is difficult because the underlying concept has several legal or economic definitions. A simple criterion is considered in order to categorise CSPs where farmers are compensated in order to reduce a negative externality. Indeed, these policies are not compatible with the polluter-pays principle.

Finally, in relation to fairness, policies leading to an increase in public access to the countryside are distinguished from other policies.

3.2.3. Analysis of criteria of special interest[1]

A series of 70 criteria have been precisely defined and filled for a selection of 143 CSPs covering the eight countries (see the Annexe). Improvement in water quality is a core objective of the majority of the CSPs reviewed. However a limited but significant number of those CSPs target the reduction of water abstraction or the control of erosion. Bio-diversity as well as water protection are a crucial objective of these policies. In comparison, human health, improvements in landscape, fire protection or animal welfare are targeted by more specifically oriented CSPs. A number of CSPs are not targeted by a precise zoning referring to geographical, ecological or agronomic features. These are policies which apply to large areas offering farmers a small compensation. This point can be linked with the transaction costs issue (see Chapter 4).

The majority of CSPs favour the provision of a good which is already marketed (such as AOC or green tourism) or could be marketed (recreational hunting for example). Farm commodity and EGS are substitutes for one-half of CSPs and complements for 38%. More than 40% favour employment locally, since they need more labour. However the effect on total employment is likely to be limited.

The control of policies implemented is generally possible but for a great number of them ease and reliability are questionable. The mode of operation does not pose specific problems. Most policies rely on an individual agreement, property rights are rented, the common policy instrument belonging to the economic or financial category. A number of CSPs are inconsistent with some policy principles because they are not decoupled or farmers are paid to reduce negative externalities.

Parametric equations have been considered to describe relationships between criteria more thoroughly. Some criteria relevant for policy-making have been selected as the dependent variables in models where the other criteria are the explanatory variables. The basic principle of the methodology is quite simple. According to the nature of the dependent criterion two econometric techniques are used. Where it is a continuous variable, a simple variance analysis is run. For

---

[1] This section has been prepared with Alain Coppens and Guido Van Huylenbroeck.

discrete variables a logistic model is applied. For both models (analysis of variance and logistics procedure) the stepwise technique is used to select the most significant explanatory criteria. This methodology requires a number of comments.

A high level of compensation per hectare is likely to be associated with CSPs requiring specific farming practices which involve a ban on nitrogen fertilisation and a limitation of pesticides. They are labour intensive. These policies are administered at a local or regional level and non-complying farmers are required to pay the money back. They favour specific features such as landscape improvements and protection of flora. In contrast, a modest level of compensation per hectare is connected with an agronomic zoning, the support of extensification with restrictions on grazing density. The underlying policies aim at the provision of quality products as well as environmental goods which support the promotion of green tourism. Policies are administered at a national scale and are likely to cover large areas. This is consistent because policies implemented on a large scale do not usually require a drastic change in farming practices as they only support the provision of existing amenities. CSPs whose scope of implementation is more limited target flora protection and scenic improvements. In addition, it is valuable to note that the highest administrative costs concern CSPs which are locally administered and applied on a small scale to areas designated in terms of ecological features (ecological zoning).

There are CSPs for which the compensation settled seems clearly high compared with the efforts required. The probability of over-compensation is significantly smaller for CSPs which are administered at a national level and are compatible with the polluter-pays principle. Non-complying farmers are required to pay the money back. The occurrence of over--compensation increases when environmental education is targeted. Otherwise over--compensation is associated with policies implemented in mountainous areas. Compensation received can represent a significant share of total farm sales: one-half of the reviewed CSPs provide a compensation over five per cent of total farm sales. This is linked with policies which are clearly incompatible with the polluter-pays principle and do not impose a specific technology for EGS provision. They usually target a limitation of nitrogen fertilisation or a ban on pesticides but do not restrict grazing density.

The probability of being incompatible with the polluter-pays principle is significantly higher for CSPs whose objective is to reduce a negative externality or to favour the conversion from intensive to extensive farming systems. These are likely to impose a specific technology for EGS provision. In contrast CSPs which are based on ecological zoning, which target wildlife protection, are likely to be compatible with the polluter-pays principle. A number of the CSPs are expected to induce a permanent provision of benefits. These sustainable CSPs promote a positive attitude towards the environment and use contracts over five years. Their speed of implementation is progressive and they do lead to the adoption of new technologies. The crucial role of education to promote new farming methods, in a progressive way in order to achieve sustainability, is emphasised. Policies whose objective is erosion reduction are considered to be sustainable. This is consistent because they often involve substantial changes in land use (plantation of hedges, conversion of arable land to grassland). Ecological zoning would not result in sustainable changes as it does not involve a drastic modification of farming practices. Here the provision of EGS is highly dependent on the payment of compensation because CSPs do not lead to the adoption of another technology. The same comment is valid for CSPs favouring extensification as they do not lead to a permanent change of farming practices. The diminution of the output maybe transitory and it is paid for during the period for which the policy is applied. It is therefore expected to stop once payments are no longer provided. Moreover, policies having a positive effect on animal production are likely to be non-sustainable. This may be explained by the support of extensive farming systems that would not be able to survive without compensation.

3.3. Establishing a typology of the CSPs

3.3.1. Principles of the methodology

A number of methods, based on a geometric viewpoint, have been considered to deal with huge data sets. They deal primarily with two-way data. This means that the data consist of a two-way array or matrix of values where the rows are associated with a set of observations, and the columns are associated with a set of variables. Here the observations are the CSPs and the variables the criteria. Among them principal components analysis and correspondence analysis are particularly relevant:

- to reveal relationships among the variables, among the observations and between the variables and the observations by dimension reduction;
- to suggest what the basic underlying variables might be.

Dimension reduction is the most important of these issues. The idea is to portray the data set by a series of projections which minimise the loss of information. So a trade-off similar to the compromise a geographer makes by projecting the earth onto a map must be found. Dimension reduction in the observations means representing the observations by a smaller number of co-ordinates. This greatly simplifies the task of studying the relationships by scatter plots or otherwise. Historically, factor analysis has given much greater attention to dimension reduction of the variables, and principal components analysis has tended to emphasise the observations. In any given application there may be good reason to emphasise one aspect or the other, but there is no intrinsic reason to continue this imbalance of emphasis. One should recognise the symmetry and utility of both aspects in order to study the relationships among the CSPs, the criteria or both. For these reasons the following section provides a unified treatment of the dimension reduction issue.

Dimension reduction

Let us consider a matrix of data:

$$X = (x_{ij}) \; i = 1, \ldots n; \; j = 1, \ldots p.$$

Where i denotes the individuals and j the variables. This matrix can be viewed as being a set of n row-vectors x_i of dimension p, thus the data set consists of n points in the space $\Re^p$. The general objective of all factor analyses is dimension reduction, which means projecting this data set in a space of lower dimension. If it was possible to visualise a p-dimension space such methods would not be useful. Since the projection distorts the data set, a measure of distortion is needed in order to assess the quality of the portrayal in $\Re$ or $\Re^2$. The projection procedure depends on the distances which have been selected but all its general characteristics can be shown with the Euclidean distance. The distance between two points i and i' is given by the usual formula:

$$d^2(i, i') = \sum_{j=1}^{p}(x_{ij} - x_{i'j})^2$$

Let us assume that all data have been centred so that the centre of gravity of the data set is equal to the origin O in $\Re^p$. The total inertia of the data set is then given by:

$$I_0 = \sum_{i=1}^{n} d^2(i, O)$$

This formula can be modified in order to take into account weights attributed to the individuals.

Now suppose that x_i is projected on the axis Δu_1 whose unitary-vector is u_1. The length c_{i1} of the projection equals the scalar product of the two vectors x_1 and u_1. Therefore the inertia of the projection of the data set on the axis Δu_1 is:

$$I_1 = {}^t C_1 C_1$$

where C_1 denotes the column-vector whose generic element equals c_{i1} and is the transposed vector of ${}^t C_1$. So the mathematical problem is to find the axis such that the inertia I_1 is maximised. The corresponding function follows:

$$\underset{u_1}{Max} \; {}^t C_1 C_1 \text{ subject to } {}^t u_1 u_1 = 1$$

Using the Lagrangian the first order condition for a constrained maximum is easily derived:

$${}^t X X u_1 = \lambda_1 u_1$$

where ${}^t X$ denotes the transposed matrix of X, and ${}^t XX$ is a positive definite matrix of order p. Thus u_1 is an eigenvector associated with the eigenvalue λ_1 of ${}^t XX$ which is called the matrix of inertia. It is easily shown that:

$$\lambda_1 = I_1$$

so λ_1 is the biggest eigenvalue of ${}^t XX$ and then u_1 is the corresponding unitary eigenvector. u_1 is the first factor. The same procedure can be applied to determine the second factor which is associated with the second eigenvalue of ${}^t XX$. There is a second constraint in the method, that u_1 and u_2 are orthogonal.

The total inertia of the data set is:

$$I_0 = trace \, ({}^t XX)$$

Inertia being additive, it can be shown that:

$$I_0 = \sum_{j=1}^{p} \lambda_j = \sum_{j=1}^{p} I_j$$

where λ_j is the jth eigenvalue of ${}^t XX$, and:

$$\lambda_i = I_j \quad \forall j$$

and eigenvalues are ordered:

$$\lambda_1 > \lambda_2 > \ldots > \lambda_p$$

Since I_j equals the inertia of the projection of the data set on the jth axis (or factor), λ_j is the inertia explained by this axis. The p axes are orthogonal. Each individual is plotted and represented in this new p-dimension basis. Using linear algebra techniques, a dual relationship between $\mathcal{R}^p$ and $\mathcal{R}^n$ can be proved. This implies that the variables can also be plotted in the same vector-basis and therefore that the distances between individuals and variables can be interpreted in the same way as distances between individuals or between variables.

Practical guidelines

The rate of inertia is often used to measure the share of total inertia explained by each factor. It is given by:

$$\tau_j = \frac{I_j}{I_0} = \lambda_j / \sum_{j=1}^{p} \lambda_j$$

Each factor is a linear combination of row-vectors (observations) so in order to interpret the axes it is crucial to consider their relative weights. The contribution is a crucial concept which offers a powerful tool. First of all recall that:

$$I_j = \sum_{i=1}^{n} c_{ij}^2$$

then the contribution to factor j of i is defined by:

$$CTR_j(i) = c_{ij}^2 / I_j$$

Due to duality between variables and observations, the contribution of each variable is defined on the same line. In order to give a consistent interpretation of the different axes, attention must be paid to observations and variables having the biggest contributions.

In order to highlight the analysis, supplementary variables or observations, named illustrative variables or illustrative observations, are often considered. They do not enter into the computation of eigenvalues and eigenvectors. However they are plotted in the new orthogonal basis. The quality of the projection is an important issue. It is assessed by the angles between vectors i, j and the axes. The squared-cosines of the angles are used instead of the angles themselves but this change does not affect the quality of the projections. The squared-cosine is also computed for the active variables and observations which are taken into account to diagonalise the matrix of inertia. But here there is a direct relationship with contributions and examining angles do not really improve the interpretation of the factors.

Principal components analysis is relevant to deal with continuous variables. But it does not work for qualitative variables. Indeed the topology of the data set is not correctly described with an Euclidean distance. Therefore other concepts of distance have to be considered to deal with qualitative variables. The chi-squared distance is well adapted to dummy variables and the correspondence analysis method is based on the underlying topology (Hill, 1974). But our data set is more complex since there is a combination of quantitative and qualitative variables. Qualitative variables do not pose any problem per se because they can always be converted into a set of dummy variables without loss of information. This procedure can also be applied to continuous variables but it leads to a loss of information. Another limitation stems from the fact that all observations and variables have equal weights. Two policies having different scopes, for example a measure restricted to a small area and a measure applied on a national scale, will equally influence the computer outcome. On the other hand, a first policy whose main objective is to promote fauna protection and a second one whose secondary objective is also fauna protection cannot be distinguished according to this criterion. A consistent way to overcome this problem is to introduce supplementary indicators since the probability of finding policies with the same profile is expected to decrease as the number of criteria increases. Despite these limitations, it is probably the only practical way to deal with a huge set of variables of different categories. This methodology named multiple correspondence analysis has been applied and the chi-squared distance used. Hierarchical classification methods have been also used in order to check the typology which was obtained.

Table 3.1
Factor description

Rank	Rate of inertia (%)	Heading
1st	16	Ecological zoning, bio-diversity conservation and green tourism
2nd	12	Organic farming and AOC
3rd	10	Control of chemical inputs: banning vs limitation
4th	8	Control of chemical inputs: limitation vs no-restriction
5th	7	Policy scope
6th	6	Targeted policies

3.3.2. Factor selection

The relevant typology of CSPs, which is the final outcome of the above procedure, results from a trade-off between statistical tests and economic arguments. Indeed factors and groups of policies must be meaningful and therefore their economic significance is of major interest. Otherwise, the objective is to achieve a robust typology of the CSPs with respect to the coding procedure and the selection of active variables. Some criteria dealing with sustainability and political acceptability are to some extent subject to value judgement and were considered as illustrative variables.

Finally 18 indicators, 11 referring to environmental effectiveness and seven to economic efficiency have been selected. Others are introduced in the analyses for illustrative purposes. Factor selection results from both theoretical and empirical considerations. But a definitive test lies in the researcher's ability to interpret factors and to extract meaningful information. As a conclusion six factors are selected (Table 3.1). Globally they explain 59% of total inertia, a share which is quite high for this type of data set.

The first factor clusters on the right side of the axis eight CSPs which promote the conservation of bio-diversity, cultural heritage and landscape. They usually encourage green tourism so their output can be marketed. Moreover they are local policies which are targeted through ecological zoning. Normally there is a ban on nitrogen and pesticides. These policies are expected to lead to an increase in the labour force as well as public access, and are compatible with the polluter-pays principle. The average Greek policy belongs to this cluster which is statistically very robust.

The second factor categorises a series of policies which promote new technologies favouring extensive farming and the conversion from intensive farming to extensive farming. These technologies are characterised by a complementary relationship and are labour demanding. Indeed, EGS are a production factor or an incidental by-product. Commodities with specific features such as AOC wines are concerned by these policies whose outputs are therefore marketed and the compensation the farmers received are consequently incompatible with the polluter-pays principle. Specific requirements involve the prohibition of mineral fertilisers and of pesticides. They are expected to favour animal welfare and to provide commodities with positive effects on human health. Aggregate impact on farm supply is revealed to be negative especially for fruit, olive oil, wine and arable crops. Sixteen CSPs disseminated across the different countries have a significant contribution to this factor. They include ten CSPs which promote organic farming and three AOC.

The third factor ranks the CSPs according to the strength of the constraints put on the use of fertilisers and other contaminants. First of all there is a discrimination between policies which ban the use of fertilisers and pesticides, and policies for which prohibition concerns only purchased inputs. For the latter, manure spreading is allowed. CSPs are significantly ordered along the axis. On the left, there is a cluster with nine German CSPs. It is opposed to a second one, located on the right side, with eight CSPs distributed in France, Italy and Greece. German CSPs were implemented quite rapidly, are transparent and their control is easy. They are based on a trade-off between the provision of farm commodities and the provision of a public good. This public good is not expected to be provided after the end of the contract. Policies in the second cluster are characterised by a very different technology, since EGS and commodities are complementary. Moreover with respect to the mode of operation, this category can be opposed to the previous one, since there is a lack of transparency, control is not easy and not reliable. To some extent these policies target fire protection and the conservation of cultural heritage. Finally their output could be marketed.

The fourth factor discriminates between policies basically with respect to water protection. Seven policies are located on the left of the axis, which is the average for Greece. There are prescriptions to limit water withdrawals and to the use of polluting inputs whereas no restriction occurs otherwise. Farmers are paid to reduce a negative externality and receive a compensation over five per cent of total sales. These policies are not compatible with the polluter-pays principle. Six policies are on the right side of the axis. They provide an output which is already marketed and are consistent with policy principles. So policies are contrasted with respect to a series of criteria describing incentiveness, enforcement, decoupling and equity.

The fifth factor discriminates between policies according to bio-diversity, the level of compensation, the share of administrative costs and the level at which they are implemented. Two clusters can be defined regarding this combination of criteria. Six CSPs are included in the first one, on the left side of the axis. It refers to generic policies which apply to the whole farm and promote improvements in fauna as well as flora, favour animal welfare and fire protection. Prescriptions are not transparent and some of them use persuasion as an instrument. These policies are implemented over a large area but on the other hand both compensation and administrative costs are low. The average French policy belongs to this cluster. There are seven CSPs in the second cluster where the Italian average is found. These policies provide a good which is already marketed. Usually they are regionally implemented and apply to a limited area. These policies are expected to significantly increase aggregate supply for both crops and Mediterranean products.

According to the sixth factor, a cluster including ten CSPs is defined. These policies target specific features such as the reduction of erosion, fire protection or the limitation of water withdrawal. Fertilisers are often banned. Moreover they are geographically zoned in favour of mountainous areas or rely on agronomic zoning. There are a number of cases in Austria, Greece and Italy. Globally the implementation of these has been rapid.

3.4. Graphic representation of the typology

3.4.1. Definition of homogeneous groups of CSPs

The typology is directly derived from the CSP co-ordinates on the six axes. Each group is defined as a cluster of CSPs around a kernel (or core). The kernel of a specific group includes the CSPs whose contributions to the group are the highest. Moreover these policies have similar profiles whereas the average profiles of the kernels are significantly different. In other words kernels are characterised with a greater degree of similarity than the whole groups and kernels are more clearly

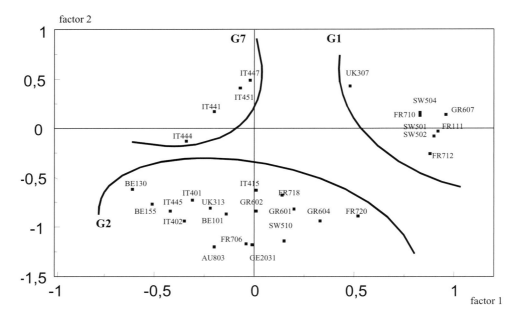

Fig. 3.2. **Classification of G1, G2 and G7 on the "output map".**

differentiated from each other than the groups. In order to clearly describe the groups of policies, this section concentrates on their kernels. Seven groups named G1 to G7 emerge. Their kernels include 56 CSPs. The number of policies considered ranges from four in the kernels of G5 and G7 to 17 in the kernel of G2. The kernel of G3 only includes German policies and the one of G7 Italian ones. Other kernels combine CSPs from at least three countries.

A clear discrimination among kernels is made with respect to input management requirements put on the various CSPs. G3 and G1 are characterised by the ban of nitrogen or other contaminants, whereas there is only a limitation in G2, G4 and G6. Otherwise there is no explicit prescription put on polluting inputs in G5 and G7. Second, the kernels are further distinguished by the nature of the EGS expected to be produced and the degree of technical complementarity between EGS and marketed goods or services. Three maps given below, are sufficient to provide a consistent graphic representation of the kernels of the groups. The classification of all CSPs is given in the Annexe of this book.

3.4.2. Representation of G1, G2 and G7 on the "output map"

Figure 3.2 emphasises the output perspective, according to which three groups, G1, G2 and G7 are clearly identified. These groups can be ordered with respect to the type of outputs which are promoted. They range from the provision of local public goods (G1) to the production of quality products (G2) and the reduction of negative externalities with no zoning of any kind (G7). CSPs included in G1 are compatible with the polluter-pays principle while the others are not.

Description of G1: Production of different kinds of local public goods by the banning of polluting inputs in ecologically defined zones under standard agreements

The ERGSs, ranging from bio-diversity to landscape beauty and cultural heritage, are simultaneously provided by each measure. Usually, they are directly or indirectly marketable, especially through green tourism. These CSPs are locally or nationally administered. This group is rather heterogeneous in terms of area, uptake and compensation level (average of 174 ECU/ha), but the share of administrative costs in total costs is usually higher than 10% and the growth of uptake is rather slow. Most of the CSPs appear compatible with the polluter-pays principle. They induce extensive and labour intensive farming practices, but they do not encourage innovation. The English ESAs and French local programmes are typical examples of this group.

Description of G2: labelled products, organic farming and AOC

Expected ERGSs are complementary with the production of high quality and healthy agricultural goods. Polluting inputs are banned or strictly limited. This group includes organic farming conversion and AOC. This is the only group where market instruments are used. Measures in this core are less transparent and they involve a less easy control than policies in other groups. They are more often regionally administered. This group has an average profile in terms of area, uptake, compensation level (average of 138 ECU/ha when different from zero) and administrative costs. Most of the CSPs favour sustainability and are polluter-pays principle incompatible, although they often have a negative impact on agricultural production in quantitative terms. The required farming practices ensuring the product quality and the co-production of EGSs are land and labour intensive and often innovative.

Description of G7: Heterogeneous group of policies of limited scope

The characterisation of this group is delicate because its kernel is very specific and homogeneous while the group itself gathers a wide range of CSPs. The kernel is constituted by Italian measures aiming at local endogenous development by enhancing rural tourism or the marketing of local agricultural quality products. The whole group is heterogeneous since it includes CSPs whose objectives range from the reduction of farming negative externality (including erosion) to the protection of cultural heritage. There is no zoning at all. Compared to the other groups, G7 has a smaller share of CSPs improving fauna, flora and aesthetic outputs. Although most of these measures do not directly restrict the use of nitrogen and other contaminants, farming practices are often explicitly constrained. For most of the other criteria, G7 has an average profile, but is characterised by a rather low compensation per hectare (average of 107 ECU/ha when different from zero) and the smallest rate of overcompensating CSPs. EGSs are rather substitutes of agricultural products and rarely marketable. Many measures are considered to be sustainable. Most of them are neither labour intensive, nor provoke extensification and give no incentive to innovation.

3.4.3. Representation of G3, G4 and G5 on the "input map"

With fig. 3.3 the focus is on the control of inputs. Along the horizontal axis, a shift from the right to the left is associated with a shift from policies limiting the use of polluting inputs to policies prohibiting it. Along the vertical axis, there is a division between policies which limit inputs and policies which put no restriction on them. Three groups, G3, G4 and G5 are then identified.

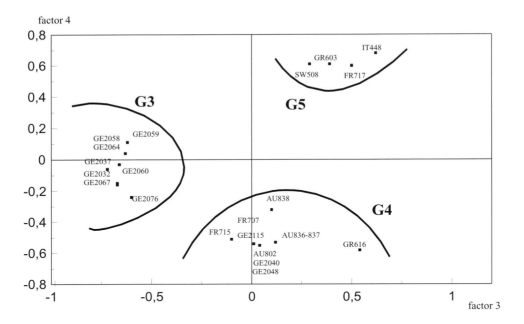

Fig. 3.3. **Classification of G3, G4 and G5 policies on the "input map".**

The corresponding clusters define a triangle. With respect to input limitation these groups can be ranked as following: G3 < G4 < G5, this reveals the so called Gutman effect and an underlying linear relationship.

Description of G3: Production of bio-diversity by the banning of polluting inputs

The EGSs are pure public goods, substitutes of agricultural goods and are not directly or indirectly marketable. They only have existence value. All included measures are transparent and locally administered, control being easy and reliable. The area concerned is limited, and the compensation per hectare is high (average of 300 ECU/ha). The uptake growth is slow. Although the compensation per farm usually exceeds five per cent of farm sales, the CSPs appear rather compatible with the polluter-pays principle. They are not labour intensive, they do not favour either innovation or sustainability.

Description of G4: Reduction of negative farming externalities

These CSPs are based on the limitation of polluting input use, livestock density and the application of a code of good farming practices. They often target the whole farm. The control is neither easy nor reliable. They are usually locally administered, never nationally. Often covering a rather large area, most of these CSPs are characterised by a high level of compensation (average of 241 ECU/ha). Their uptake growth is often fast. Farmers are usually overcompensated and get more than five per cent of their sales through this channel. This group has the highest rate of polluter-pays principle incompatible measures. Most of these measures lead to a decrease in agricultural production. Innovative farming practices are often required.

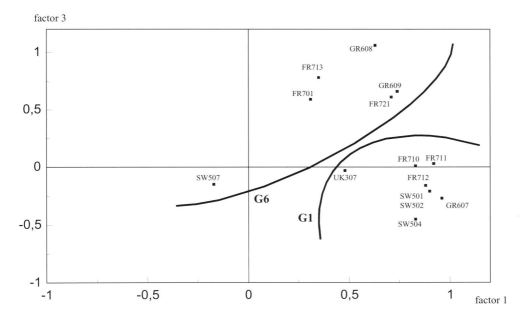

Fig. 3.4. **Discrimination between G1 and G6 on an "output-input" map**

Description of G5: Support of specific features of the rural landscape

This includes features like hedges, trees or threatened breeds which may be endangered or abandoned by intensive farming practices. These CSPs usually have a positive impact on erosion control without restricting farming practices. The EGSs produced are usually intentional or main products (substitutes of agricultural goods) concerning bio-diversity, landscape beauty or cultural heritage, always marketed, or marketable by green tourism. The participants are not compensated for farm production loss due to restrictions but for the EGS production itself under standard or individual agreements. These measures are usually transparent and easy to control, but sometimes no repayment is stipulated. The compensation per hectare equivalent is usually low (average of 161 ECU/ha). The uptake is often large in terms of participants and equivalent area. Even if total compensation per measure is high, the compensation per farm often remains under five per cent of sales. These measures are usually considered PPP compatible. They have no impact on agricultural production. They can require labour intensive practices but no innovative practices.

3.4.4. Representation of G6 on an "output-input" map

Description of G6: Production of different types of public goods by the limitation of polluting inputs or livestock density

Reduction of erosion and fire protection are often targeted. Water use can also be restricted. The whole farm is often concerned. The ERGSs, ranged from bio-diversity to landscape beauty and cultural heritage, are simultaneously provided by each measure. They often are directly or indirectly marketable. Compared to other groups, the ERGSs are often farm production factors. Only half of these measures are transparent. The control is both difficult and reliable. They are locally or nationally administered. Each measure concerns few farms but a rather large area. Most of these CSP are characterised by a high total compensation. The uptake growth is often rapid. The share of administrative cost in total cost are under 10% for half of the measures, but higher than 50% for one third. Most of the measures are possibly polluter-pays principle compatible, and do not appear sustainable. They have no impact on agricultural production. Their production can require innovative or labour intensive farming practices. Compared with G1, the expected environmental outputs are similar (axis 1) but the input restrictions are weaker in this group (axis 3). The average compensation per hectare is 165 ECU for the CSPs of G3, which is very close to G1 (average of 174 ECU/ha). As the restrictions are weaker, the relatively high compensation per hectare may explain the higher uptake than in group G1.

3.5. Concluding comments

The distribution of CSPs according to countries and groups (Table 3.2) summarises the outcome of the typology. The preparation of the data base resulted from a learning process in which the different research teams participated. The information has therefore been collected in a consistent way across countries. However, it shows a country bias which mainly results from the way in which the Regulation 2078/92 is implemented in the eight STEWPOL countries. But it is also partly due to the way in which each partner selected the sample of policies. So differences between countries must be carefully considered. Anyway there is a close relationship between G3 and Germany, G4 and Greece, G7 and Italy, and finally G1 and the cluster made by Sweden and the United Kingdom. Other countries have an average profile.

Figure 3.5 gives an overview of the seven groups on a two-dimensional diagram and clarifies several points. The two axes incorporate the input and the output perspectives together. The horizontal axis refers to the limitations put on inputs whereas the vertical one emphasises the underlying technology. G1 and G3 gather CSPs which require the ban of nitrogen or other contaminants. In G2, polluting inputs are banned or at least strictly limited. In G4 and G6, quantitative limitations are required. In G5 and G7, there are no explicit quantitative restrictions on nitrogen or other contaminants even if some CSPs in G7 may require specific farming practices leading to actual input restrictions. The vertical axis refers to the complementarity or substitution

Table 3.2
Distribution of CSPs according to countries and groups

	G1	G2	G3	G4	G5	G6	G7	Total
AU	3	2	2	5	1	1	5	19
BE		3	3	2	2		5	15
FR	4	3	1	5	2	4	2	21
GE	3	1	10	6	2		8	30
GR	1	5		11	1	2		20
IT		4	1		2	2	14	23
SW	6	1			1	1	1	10
UK	5	1		2			2	10
Total	**22**	**20**	**17**	**31**	**11**	**10**	**37**	**148**

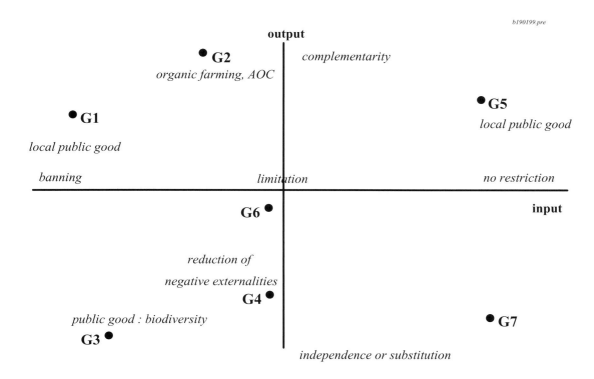

Fig. 3.5. **Characterisation of the groups.**

relationships between the expected EGS of the CSPs and marketed goods (cf. Chapter 2). The degree of complementarity is higher toward the upper part of the diagram whereas it is lower toward the lower. In G5, the EGS itself is considered marketed because compensation payments are proportional to the EGS production of the participants. In G2, the support to organic farming and farm product labelling is justified by the supposed complementarity between these types of farm products and the expected EGS. On the opposite side in G3, the expected EGS are pure public goods, and are moreover substitutes or independent of marketed farm products. The CSPs in G4 and some CSPs belonging to G7 aim at the reduction of farming negative externalities, leading to the decrease of agricultural production. In the intermediate situation, the expected EGS of the CSPs in G1 and G6 are local public goods favouring the production of marketed goods and services through green tourism for instance.

The stronger the restrictions on farming practices, the higher the compensation per hectare and the share of administrative costs in total costs are. However, this statement does not apply to G2 which gathers CSPs using market instruments such as recognised labels. The more the payments overcompensate the required efforts, the larger and the faster is the uptake. This situation most takes place in G6, G4 which include the CSPs with the highest total exchequer cost, although their shares of administrative costs in total costs are rather low. Hence, the crucial issue of environmental benefit measurement arises. No doubt the benefits per hectare are higher in G1 or G3 than in G4, G6 or G2. But we do not have any evidence concerning the balance between the total benefits and the exchequer costs for the different types of CSPs.

The CSPs of G2 and G4 are those which are expected to decrease agricultural production the most. The situation of G4 is awkward because the related CSPs are both incompatible with the polluter-pays principle and among the most expensive. With respect to the CSPs of G2 two main issues emerge, especially when financial supports are involved. First, the expected EGS not being explicitly defined makes it difficult to disentangle amenities which are remunerated through higher market prices from amenities which justify the related financial supports. Second, these CSPs would be compatible with international trade principles only if the financial supports are clearly decoupled from the labelled production.

Stronger evidence could be derived from this typology if a set of measured environmental indicators would be included in order to characterise the different groups. The typology itself could be improved by unifying the collection of the useful indicators. In spite of the efforts spent in an unambiguous definition of the current indicator set, national biases might remain because of their interpretation by the national experts who collected the data. Nevertheless, any new CSP, characterised by its objectives and policy instruments, is easily classified according to this typology which provides expectations about its uptake, cost and compatibility with policy principles.

References

Abler, D. G. and Shortle, J. S. (1990) Environmental and farm commodity policy linkages in the US and the EC, *European Review of Agricultural Economics*, 19, 197–217.

Adger, N. and Whitby, M. (1991) Accounting for the impact of agriculture and forestry on environmental quality, *European Economic Review*, 35, 629–641.

Bouwes, N. W. and Schneider R. (1979) Procedures in estimating benefits of water quality change, *American Journal of Agricultural Economics*, 61, 535–539.

CBO (1994) *Greening the national accounts*. Congressional Budget Office Papers, Washington D.C.

CBS (1992) *Natural resources and the environment 1991*. Central Bureau of Statistics of Norway, Rapporter 92/1A.

Hanley, N. (Ed.) (1991) *Farming and the countryside: an economic analysis of external costs and benefits*. CAB International, Wallingford.

Hill, M. O. (1974) Correspondence analysis: a neglected multivariate method, *Applied Statistics*, 23, 340–354.

Just, R. E. and Antle, J. M. (1990) Interactions between agricultural and environmental policies: a conceptual framework, *American Economic Review*, 80, 197–202.

Just, R. E. and Bockstael, N. (Eds) (1991) *Commodity and resource policies in agricultural systems*. Springer Verlag, Berlin.

Lipsey, R. G. and Lancaster, K. J. (1956) The general theory of second best, *Review of Economics Studies*, 24, 11–32.

Moxey, A. Whitby, M. and Lowe, P. (1998) *Environmental indicators for a reformed CAP: monitoring and evaluating policies in agriculture*. Centre for Rural Economy, University of Newcastle upon Tyne, UK.

Parris, K. (1999) Environmental indicators for agriculture: overview in OECD countries. In: F. Brouwer and B. Crabtree (Eds) *Environmental indicators and agricultural policy*. CAB International, Wallingford.

Vail, D. L., Hasund, K. P. and Drake, L. (1994) *The greening of agricultural policy in industrial societies*. Cornell University Press, Ithaca and London.

Whitby, M. (1996) Losers and gainers from rural policies. In: P. Allanson and M. Whitby (Eds), *The rural economy and the British countryside*. Earthscan, London.

G. Van Huylenbroeck and M. Whitby
Countryside Stewardship: Farmers, Policies and Markets
© 1999 Elsevier Science Ltd. All rights reserved

Chapter 4

The invisible costs of scheme implementation and administration

Katherine Falconer and Martin Whitby

Abstract—The crucial role of government administration, and its related resource use is currently understated systematically in economic analyses of the provision of public goods by procurement from landowners. This chapter centres on the organisational costs of economic systems, especially those incurred in the public sector in relation to agri-environmental policies. Scheme administration entails real resource costs, raising economic and financial issues, particularly in terms of tax-payer value-for-money and scheme design to procure agri-environmental goods efficiently. Identification of the determinants of transactions costs is essential in considering how scheme administration costs might be reduced while still delivering policy benefits. Non-trivial policy-related transactions costs, i.e. administrative costs, stem largely from factors such as the heterogeneity of producers and the asymmetry of information between land-holders and public agencies. Analysis of an unique quantitative data-set of administrative cost estimations for 37 case-study schemes in eight European member states suggests that there are strong links between administrative costs and the time profile of scheme participation. Different activities are needed at different stages of the scheme's life; generally, agreement maintenance is much less expensive than agreement set-up, contributing to a fall in costs with time. While it is currently too early to tell, there may also be downwards pressure on costs over time from economies of scale and the economising effects of experience. Comparisons of data on agricultural support and agri-environmental policy suggest that administrative costs may become even more important as environmental objectives receive greater priority in policy-making.

4.1. Introduction

Agricultural support in the EU is thought to have led to rising levels of land use intensity in recent decades, threatening widely-valued characteristics of the countryside. A common belief now is that environmental and amenity public goods are no longer produced jointly with market commodities.

Hence policies have developed seeking to encourage the provision of agri-environmental public goods, particularly through their direct procurement from farmers through fixed-term land management contracts.

At present, the main financial cost component for many agri-environmental schemes across the EU – particularly, those implemented under Regulation 2078/92 and Regulation 746/96 – relates to compensation payments to landholders. However, the *gross* public exchequer costs encompass both payments to farmers and the organisational costs of schemes.[1] If scheme operational costs are a potentially substantial component of the total policy-related costs borne by the public exchequer, consideration of them should be included as an element in policy decision-making. Furthermore, there may be barriers to participation in voluntary agri-environmental schemes if farmers incur significant transactions costs of their own, for example, related to making initial enquiries about scheme participation. The existence of constraints on participation is important insofar as it may jeopardise the achievement of policy objectives.

Whereas only some agri-environmental schemes involve payments to farmers, all schemes, regardless of their type, cause administrative costs to be incurred by the implementing agency. In economic terms, compensation payments are transfers only, whereas the administration of these payments incurs real resource costs, which are of direct relevance to resource allocation. The costs of contracting reduce welfare: partly by absorbing resources directly and partly by suppressing exchanges that otherwise would have been mutually beneficial. However, at present, the levels, structure and incidence of the organisational costs of policy implementation are typically undocumented, and it is unusual to find policy evaluations in the agri-environmental literature which include them, despite widespread recognition of their importance (for example, Stavins, 1993; Latacz-Lohman and Van der Hamsvoort, 1998).

In contrast there has been considerable attention, in economic policy analysis, to the opportunity costs of agri-environmental policy at farm-level, in terms of lost production. Most governments currently fail to report the organisational costs of scheme implementation, resulting in a low transparency of scheme costs, with the potential, thus, of failing to ensure the best value for money for tax-payers. While the existence of administrative costs does not imply government failure, there is certainly a problem of invisibility. The aim of this chapter, therefore, is to throw more light on the transactional sphere in agri-environmental policy, as an area that has received only slight empirical attention to date in the policy evaluation literature.

Resource scarcity implies that organisational costs should be optimised against the objectives of the policy. The administrative efficiency of agri-environmental schemes are of growing contemporary importance in practical policy-making discussion; see, for example, the National Audit Office report (NAO, 1997), which considered the organisational effectiveness of the UK Environmentally Sensitive Areas (ESAs). Although agri-environmental policy sphere is relatively small at present in terms of overall public expenditure, it looks set to expand in the future. Hence it is important to assess as accurately as possible the full resource use implications, including organisational aspects of resource use, of policy development under different scenarios. It would be very useful to understand and assess the determinants of policy-related transactions costs, in order to be able to set these costs against the effectiveness of policies, to improve their evaluation and ultimately to improve the value for money of public expenditure in the agri-environmental sphere. In addition, the identification of factors leading to relatively high or low administrative costs for schemes might aid reductions in costs to be made while still allowing policy goals to be achieved.

---

[1] The implementation of some schemes may also induce changes in exchanquer costs for other policies, such as agricultural commodity supports, so ideally the *net* costs of a scheme should be calculated and used in any evaluation (see Saunders 1996).

Schemes based on voluntary management agreements between the state and private producers are particularly interesting in the transactions-economics context as in effect they amount to a 'quasi-market' for agri-environmental goods (i.e., farmer/State contracting). Section 2 examines the nature of agri-environmental transactions relating to voluntary schemes; the empirical approach is developed in Section 3. Section 4 presents the results of the data analysis for almost forty schemes across eight EU member states, and Section 5 discusses the findings in their broader context. Section 6 concludes the chapter.

4.2. Transactions costs and contracting in agri-environmental policy

Organisational costs, or 'transactions' costs have not been defined precisely, to date. Cheung (1987) defined them in the broadest sense to encompass 'all those costs that cannot be conceived to exist in a 'Robinson Crusoe' economy where neither property rights nor transactions, nor any kind of economic organisation, can be found'. Arrow (1969) defined transactions costs to be the costs of running the economic system. Niehans (1971) defined them as those costs that arise not from the production of goods, but from their transfer from one agent to another. At root, essentially, are the information deficiencies faced by one or both of the transacting parties, and the costs of removing such deficiencies (see Dahlman 1979). The chief reason for their existence is the degree of heterogeneity of the characteristics of the type of commodity to be exchanged; hence, for example, the need for monitoring and compliance enforcement, given the presence of opportunism in contracting with asymmetric information.

The current situation in the EU is one of perceived under-supply of agri-environmental public goods. It would be useful, therefore, to understand why dis-equilibrium persists between the supply and demand of agri-environmental goods, and then to ask how provision could be increased. A starting point for analysis lies in the breakdown of classical contracting for agri-environmental goods (see Coase, 1960). Public goods and externalities exist when the private economy lacks incentives to set up a market for a good (i.e., the costs of change exceed the anticipated gains), and when the non-existence of this market results in a Pareto sub-optimal resource allocation. Transactions costs often present barriers to the efficient resolution of conflict through the market mechanism.

In the agri-environmental sphere, free-market exchange between individuals is often prohibitively expensive, for example, given the high costs of co-ordination where the agri-environmental goods and services which are demanded (such as traditional pastoral landscapes) are produced in non-separable ways by different land-owners and are consumed by different individuals. Information costs are high given characteristics such as the variable, often highly location-specific, nature of agricultural production technology and opportunity costs, and the variable natural heritage value (and potential value) of any parcel of land; the complexity of the management processes and components required for both agricultural and natural heritage production and the low observability of most management activities. Information asymmetry gives rise to significant problems. Variability in the attitudes (objective functions) of each individual land-owner means that there will be different levels of opportunism against which the 'buyer' of agri-environmental goods must safeguard. The influence of stochastic environmental factors, such as weather conditions, also means that there is an inherent level of uncertainty to agri-environmental production.

These characteristics pose significant hurdles to private market contracting with regard to agri-environmental goods and services, resulting frequently in unresolved resource use conflicts. In some cases, these conflicts will motivate policy development to improve resource allocation. However, the transactions costs that inhibited free-market provision of goods will also have implications for the cost-effectiveness of policy mechanisms to procure them from the private sector, as the nature of the goods remains unchanged. The counterpart to the transactions costs of decentralised markets is administrative costs (for example, relating to policy implementation), which can similarly give rise to dead-weight and impose inertia on the initial assignment of property rights.

If transactions costs (and income effects) are zero, the form of economic organisation will not influence resource allocation: the procurement of goods and services through either firms or markets will be equally efficient (see Williamson, 1985). However, the real world is characterised by positive transactions costs: hence, designers and evaluators of economic systems should take them into account, since their omission from the decision calculus could otherwise result in sub-optimal policies. Nevertheless, while the empirical importance of transactions costs to the functioning of environmental policy is not disputed, most neo-classical economic policy analyses do not explicitly consider them (Stavins, 1993, although see McCann and Easter, 1998)[2].

Governments in many countries (and especially in Western Europe) have tried to stimulate agri-environmental goods production through the development of administratively-run markets for agri-environmental goods. Compensation payments are made by the State in return for commitments from land-owners to manage the land in particular ways specified with the objective of producing environmental goods and services. The State is, in effect, a buyer, and landowners are in the position of sellers. Such mechanisms work primarily through 'collectivising' agri-environmental transactions, i.e. by reducing the search costs of buyers and sellers, facilitating transactions, and thus allowing improvements in the resource allocation to be made. If transactions costs are positive, performance will depart, possibly substantially, from the least-cost ideal. So, for example, in agri-environmental quasi-markets, less 'trading' (i.e., scheme participation) will occur between farmers and the State than would be expected from simple economic models that do not take account of organisational costs. It can be misleading to use the standard practice of comparing conventional policy instruments with a least-cost benchmark and assuming that the latter represents the performance of a market-based instrument.

Theories of economic organisation developed by Williamson (1985) can be applied to guide thinking with regard to agri-environmental policy design, and to provide some insights into the relative cost-effectiveness of different approaches to provide agri-environmental goods to guide future policy developments. Rather than considering simply the *production* of agri-environmental goods, the alternative policy structures and their relative appropriateness to provide different types of agri-environmental goods can be considered from a transactional perspective. Given the positive transactions costs of policy implementation based on voluntary schemes, the form of contract must be chosen carefully if economic efficiency is to be achieved. For economic efficiency, that scheme or mix of schemes which minimises *total* costs—i.e. scheme compliance costs (production costs or the opportunity costs of producing agri-environmental goods) *and* transactional costs—should be chosen. However, there may well be different trade-offs between these two components under different scheme organisational structures: consideration of either component in isolation may mean that the policy framework develops in a sub-optimal direction. So, the question arises: if a market is to be established for agri-environmental goods, what type of market would be best?

Williamson (1985) considered transactional attributes such as information asymmetry (related to the degree of observability of the attributes of the good or service to be provided), opportunism, and asset specificity to influence the efficiency of any given mode of economic organisation. In the agri-environmental sphere, as heterogeneity increases, in terms of both the producer opportunity costs and agri-environmental output, the search and information costs will rise. The aim is to

---

[2] Although there is a substantial theoretical environmental-economics literature on transactional aspects of policy such as information asymmetry in State-farmer contracting: see, for example, Latacz-Lohmann and van der Hamsvoort (1998), and Moxey et al. (1999).

identify the transactional system that will balance the overall costs of compensation payments against the costs of compliance monitoring and payment administration. Heterogeneity in agri-environmental compliance costs poses a considerable challenge to efficient scheme design, since such costs are generally not directly observable by regulators without incurring substantial monitoring costs. Hence, there has been a tendency, to date, to opt for the administrative ease of an undifferentiated payment mechanisms in agri-environmental contracting, even though this leads to adverse selection (Moxey and White, 1998).

Before taking the theoretical transactional analysis further, when considering policy development in the agri-environmental sphere, more information is needed on the actual incidence and magnitudes of the administrative costs of different types of procurement schemes, given how little is known of their economic significance at present. Scheme-related transactions costs have resource use implications in both the public and the private sectors, i.e. on both sides of an exchange, through constraining the type and number of exchanges each can make. For example, the main effect of transactions costs incurred by 'sellers' of agri-environmental goods is to reduce participation in a voluntary scheme, as the privately-incurred transactions costs of participation shift the supply curve upwards. Greater understanding of the costs of administering different contracting options would allow a more constructive re-thinking of policy approaches. The data-set collected under the STEWPOL project can be used to inform the future development of the theoretical framework. The rest of this chapter contributes some empirical analysis for schemes based on voluntary compensated management agreements.

4.3. Empirical administration costs estimations for agri-environmental schemes

4.3.1. Cost typologies and incidence

Like production costs, transactions costs cover a heterogeneous assortment of inputs, but there is as yet no accountancy convention for them. A starting point is the examination of the transactions that occur, and the transactions that are needed as the bare minimum for 'effective' policy operation. The number and scale of schemes based on voluntary, compensated, multi-annual management agreements with private landowners have expanded considerably following the implementation of Regulation 2078/92. Table 4.1 summarises the principal components of the transactional (administrative) costs for such procurement schemes.

It is useful to draw a distinction between transactions costs that are fixed or variable with the level of participation in the scheme. The early years of any scheme are characterised by relatively high fixed costs relating to set-up, and afterwards, other fixed costs relating to scheme evaluation and development will be incurred. These will be independent of the area entered into the scheme, or of the amount transferred. Given the existence of fixed costs, administrative economies of scale may also be important. Costs which are variable at the level of the participant will be incurred through several different transactional activities, such as record-keeping; farm mapping; conservation plan development; processing scheme application forms; processing annual payment claims where compensation is available; and farm-visits for compliance monitoring. A challenge to the efficient administrative management for voluntary schemes lies in the difficulties of predicting changes in participation levels (particularly through new recruitment each year), which are important drivers of the variable component of costs. The incidence of transactions costs falls on both the public sector and the private sector. The costs to farmers are a mixture of fixed costs (for example, in relation to acquiring and reading promotional material about schemes) and costs that are variable with parameters such as the number of hectares entered into the scheme (for example, farm

Table 4.1
Categories of transactional costs incurred in the implementation of voluntary schemes based on compensated management agreements

Main Category	Sub-Category	State agency costs		Participant costs	
		Fixed	Variable with no. of participants	Fixed	Variable (e.g. with hectares entered)
Information	– surveying of the designated area	✓			
	– designation of area and designing management prescriptions	✓			
	– re-notification/re-design of prescriptions	✓			
Contracting	– promotion of scheme to farmers	✓	✓	✓	
	– negotiation between organisation and farmer		✓	✓	✓
	– administration of contract (making payments)		✓	✓	✓
Policing	– environmental monitoring and scheme evaluation	✓			
	– enforcement of farmer compliance		✓	✓	✓

mapping and conservation auditing prior to entry). In the policy context the important costs are those which are the marginal result of possible changes. However they must be expressed as marginal costs per unit of product, which is particularly difficult to isolate from the data available. Preliminary work on panel data for 22 English ESAs from 1991–96 concluded tentatively that the cost of starting up a new agreement was £2000 and subsequent maintenance of it cost £500 per annum, and the extra cost per hectare of new agreement was £3 (Falconer and Whitby, 1999a).

While the focus of this work is on voluntary management agreement schemes, in the broader context it is important to point out that different types of policy approach are likely to have different transactional cost structures, as illustrated in Table 4.2 below.

Of interest are the factors affecting the magnitude of organisational costs of schemes. A conceptual model for voluntary schemes based on compensated management agreements is outlined in Falconer and Whitby (1999a). For example, a more complicated scheme, with more management options, is likely to raise the costs of negotiating and enforcing agreements. More stringent requirements (higher costs to land-owners in terms of foregone production) mean that for any given probability of detection, cheating is more rational, raising enforcement costs. However, farmers with more positive attitudes towards the scheme and its conservation objectives might be expected to be more co-operative, reducing the transactional costs of establishing management agreements and the costs of monitoring and enforcement. There may also be some positive spillover effects from the implementation of other, related policies; total transactions costs might not increase in a linear way with the number of additional schemes as the costs of some activities (such as the initial surveys, ecological monitoring through farm visits and so on) can be shared.

Table 4.2
Policy approaches and administrative costs

	Information / set-up / promotion	Contracting	Policing
Persuasion and Advice	✓		
Regulation	✓		
Market mechanisms (e.g. taxes)	✓		
Tradable permit schemes	✓	✓	✓
Voluntary management agreements	✓	✓	✓
Public purchase of land	✓	✓	

Alternatively, however, administration costs may be increased through the need to co-ordinate schemes and prevent double payments and so on. These are just some of the factors to consider when examining the operation of agri-environmental schemes.

This study focuses on the public sector as the first stage, given a pragmatic consideration that there are fewer 'public' agencies for which to assess hidden administrative costs, compared to the vast number of dispersed market agents. Interest here lay particularly in the marginal costs (to governments) of a policy change (for example, of extending scheme coverage by an extra hectare or by making one extra management agreement), given an existing, well-functioning broader policy framework. There is a broad spectrum of institutional costs to identify and assess, but to keep the research within manageable bounds, the scope of enquiry was focused on the direct set-up and operating costs of schemes. This focus required concentration on those government departments and agencies that administer the schemes at first hand and excluded, for example, bodies that make available the funds for payment or establish relevant legislation. In addition, unless private agents are altruistic, they will cover the transactions costs they incur through scheme participation with the compensation payments received, i.e. private transactions costs are ultimately also funded public sector (see fig. 4.1). This further justifies the focus on the direct organisational costs of schemes as these costs are definitely not included in other components of scheme exchequer costs. Hence, their explicit measurement and inclusion in the calculation of the overall costs of schemes is necessary for comprehensive evaluation.

Furthermore, as flat-rate payments are used for most of these policies, rather than individually negotiated payments which might trace out the supply curve of conservation goods, compensation payments are virtually certain to include a more or less substantial element of economic rent. This has been crudely estimated for ESAs in the UK (Whitby and Saunders, 1996).

4.3.2. Methodology

The STEWPOL project aimed to take policy cost evaluation beyond farmer compensation costs (where paid) through assessment of the magnitude of the costs of administering various agri-environmental schemes in absolute and relative terms. The objective was to produce a comparative study of the public transactions costs of countryside stewardship schemes across eight European member states, based on a number of research questions:

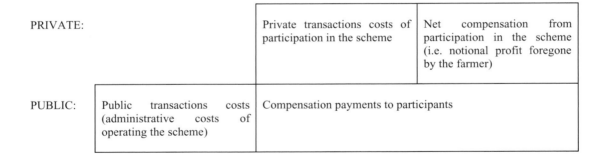

Fig. 4.1. **The relationship between public and private policy transactions costs**

1. What are the magnitudes of administrative costs in the agri-environmental policy sphere, in absolute terms and relative to other policy cost components, such as compensation payments?
2. How do administrative costs vary over schemes, within member states and over different member states?
3. What causes variation in the magnitudes, and relative importance, of administrative costs?
4. Can scheme organisational costs be reduced while still achieving policy goals?

Several schemes were selected from each of the STEWPOL countries, drawn from the wider set of rural policies that aim to reduce the negative impacts of agriculture or to stimulate positive impacts of on the countryside through land management. Details of the schemes (37 altogether) are found in the Annexe of this book, see also Falconer and Whitby (1999b, Appendix 2). A high degree of pragmatism was involved in the case-study design: schemes were selected for which data were available or could be collected, as well as on the basis of differences in design with a view to providing complementary information.

The main challenge to transactions costs analysis for different schemes across the EU lies in drawing together disparate information: for different schemes, with different objectives, implemented at different times, under different economic (and social) conditions, and within different legal and political frameworks. A broad qualification to the estimations and their comparison is that it is still early in their life for full evaluation of agri-environmental policies. Many of them are also of an experimental nature. Comparisons across schemes are not easy given the number of aspects of their design and differences in these to take into account. Another important *caveat* for analysis is that different schemes were introduced at different times.

Administrative cost data and information on scheme participation were gathered by STEWPOL researchers in each country over as many years as possible (with a range of one to eight years; the period of study was constrained by record-keeping and the time since scheme establishment). A major component of public transactions costs is the cost of administrative staff in agencies and their overhead costs. However, as little documentation of such costs is available from most governments, specific methods had to be developed to quantify them. Direct measurement approaches were applied in most member-state analyses, with estimations of the total time per year needed to carry out the necessary activities based on detailed histories of events relating to each scheme's development and operation. A standard salary rate was then applied to each staff-day. This is a blunt method of estimation, but it recognises the inherent imprecision in organisational-cost data; differentiation might otherwise give an impression of greater rigour than is feasible.

There are important issues of transactions cost definition, requiring great care at the empirical data collection stage to ensure consistency in estimations (i.e. that similar costs are measured and included). The comparability of different estimations must be assessed for valid cross-country (and cross-scheme) comparisons to be made; it is also necessary to assess the extent to which differences in costs can be explained by methodological disparities. Some differences in the components included in the administrative cost estimations for different member states were noted. For example, some studies were confined to contracting-related public costs (i.e. the costs of concluding and operating management agreements, excluding policy development costs). Other studies included the broader costs, such as initial surveying of areas to be covered by the scheme, design and re-design of the scheme following evaluation and so on. For example, the Greek and the Belgian estimates include components related to the set-up costs of policies, such as the costs of officials involved in achieving European Commission approval for schemes under Regulation 2078/92, whereas the French and English case studies did not. Overheads were also treated in different ways across the studies. The UK data on administrative costs for Regulation 2078/92 schemes included significant expenditure on environmental monitoring, with no parallel elsewhere. Nevertheless, the STEWPOL data-set provides a valuable starting point for future work. One approach would be to compare cost estimates on the basis of pure 'operations' costs to reduce the bias in comparisons, for example, by removing scheme development costs (such as the costs of seeking approval for financial support from the European Commission) and environmental monitoring costs from the analysis.

Time lags between payment applications (and conservation management actions) and actual payment transfer are another issue in time series analysis. For example, in Belgium, there were delays in making payments to participants in the pollard-willows scheme, so applications made in 1994 were not financed until 1995. Under most schemes, payments are made in arrears. This time-lag implies that indicators such as scheme administrative cost as a percentage of total scheme costs or of scheme compensatory payments will be distorted. This problem must also be viewed in the context of the incompleteness of the administrative cost estimates (compared to the more concrete figure for compensatory payments). In the absence of sufficient data to remove lags systematically, no adjustments were made.

Typical administrative wage rates vary substantially across the case-study countries, giving rise to problems for comparative economic analysis. Such variation might not, perhaps, matter if cost differences reflected differences in the calibre of the administrative staff and their inputs. To the extent that wage costs reflect the degree of training and competence of officials, expenditure may be a reasonably good indicator. However, there is no way of gauging whether or not this is the case; many economic and political factors affect wage rates and we would not expect administrative staff to be paid the same in all member states. An inherent limitation to analysis is that administrative expenditures do not necessarily reflect the quantity and quality of administrative activity, either in terms of that administration actually carried out or of that effort thought to be required for efficient running of the scheme. An alternative methodological approach is to assess levels of administrative resource use using time inputs rather than cost. This approach would enable assessment and comparison of scheme administrative resource use in terms of hours per hectare, per application, per ECU of compensation paid and so on. Only the French, German and Greek case studies detailed the time requirements estimated for the administration of each scheme, so the hours spent on particular schemes in other member states were deduced from the estimates of overall administrative costs, given typical administrative wage rates in each country for each year.

Another consideration relates to budgeting processes. The public budget setting process and constraints upon it mean that administrative inputs are unlikely to be optimal at any given time. Administrative expenditures will depend to a large degree on how agencies forecast farmer participation and administrative resource needs relative to the likely workload in each year. Prediction and budgeting are never perfect, and public budgeting processes and constraints mean that administrative inputs are unlikely to be optimal at any given time. The presence of 'noise' must be considered, such as the adverse effects on administrative costs of staff sickness, turnover and the need to train new staff, loss of expertise and so on. Inflexibility in administrative structures must be considered, such as the fact that commonly staffing adjustments are only on a yearly basis.

While the main focus of the research was placed on the public administrative costs of schemes, the underlying interest lies in overall policy-related transactional costs, in both public and private sectors. Accountancy problems must also be borne in mind, in terms of the incidence of transactions costs on the state and on scheme participants. If farmers are not altruistic; rational, profit-maximising farmers will not enter schemes unless all their costs are covered. Thus, private transactions costs will be hidden in the compensation payments, giving a downwards bias (to a currently unknown extent) in per-hectare and per-agreement figures for administrative costs. Administrative cost indicators based on the ratio of administrative costs to compensation costs are doubly biased, as the denominator is likely to include some element of the organisational costs incurred by farmers while the numerator covers public administration costs only.

Finally, it is still premature to evaluate agri-environmental policies, particularly given the experimental nature of many of them. Some have had explicit pilot phases, and most have varied over time with regard to their conservation prescriptions and in some cases with regard to their administrative frameworks. The different administrative cultures and traditions, state-market structures and so in each country, and the different stages of development of agri-environmental policy frameworks, must be borne in mind too (see Chapter 2 for more detail).

4.4. Findings from the case studies

Measurement of the generally-hidden organisational costs of schemes can give some idea of the amount by which the (direct) public costs of countryside stewardship policy are being under-estimated, for example, in academic policy evaluation studies or in official reports on scheme expenditures. An aim of this research was the presentation of more appropriate estimates of the public costs of schemes through estimation of their administrative costs and inclusion of these costs in policy evaluations.

The main finding of the case-study analyses related to the substantial under-estimation of the public costs of agri-environmental schemes, with implications for scheme evaluation, for example, in terms of value for money, and in terms of scheme development. In the agri-environmental sphere, non-trivial policy-related transactions costs, i.e. administrative costs, stem largely from factors such as the heterogeneity of producers and the inevitable asymmetry of information between land-holders and public agencies. The complexity of conservation management of different holdings means that there are few standards or blueprints for plans, and there will always be a fair degree of idiosyncrasy.

The total annual administrative costs of schemes in the mid 1990s vary greatly, depending at least partly on coverage and participation levels. In Belgium, around 20,000 ECU were spent annually on the organic aid scheme,[3] compared to annual costs of 600,000–1,500,000 ECU on the French arable conversion scheme (*Reconversion des Terres Arables*), and 900,000–1,400,000 ECU for the livestock extensification scheme (*Diminution des Chargement du Cheptel*). These figures demonstrate that potentially very large amounts are spent annually on administration, i.e. that scheme organisation is a non-trivial cost component and should feature in scheme value-for-money evaluations, although scheme scale should of course be taken into account for meaningful comparisons to be made. Farmer-contracting and activities such as routine

---

[3] The low level of costs was attributed to the requirements for participants to register with independent certification bodies, in effect transferring transactions costs from the public sector to the private sector (although to an unknown degree).

Table 4.3
Weighted average annual administrative costs for case-study agri-environmental schemes in each member state in the mid 1990s

	Average annual administration costs, ECU per hectare[1]	Average annual administration costs, ECU per participant[2]	Average annual administration costs, ECU per 100 ECU paid as compensation[3]
Austria	20.5	216.9	8.8
Belgium	58.6	388.6	63.4
France	75.6	1522.0	87.1
Germany	10.2	177.5	12.3
Greece	59.7	470.1	8.6
Italy	13.1	140.0	6.6
Sweden	9.1	190.4	11.3
UK	48.0	2445.5	47.9

[1] area-weighted; [2] participant-weighted; [3] expenditure-weighted.

payment processing seem to account for the greatest proportion of total annual administrative resource use for many schemes, although the relative shares of the different administrative cost components (information, monitoring, and general administration such as processing payments) varied across time, across country and across measures.

Scheme administrative costs were standardised by variables such as scheme participation levels, and the time that had elapsed since the scheme was first implemented. Scheme scale varied a great deal, in terms of cumulative hectares entered under each one, the number of contracts made, and the amount of compensation paid each year. Some schemes were very small: for example, the two management agreement schemes for Land Consolidation Scheme projects in Belgium had total land entries respectively of seven hectares and twenty-one hectares. Other schemes were very large, such as the French local-level (OPL) schemes, covering over 400,000 hectares; similarly, over 400,000 hectares were under conservation contracts in the English ESAs. The numbers of participants varied too, from only two farmers in the management agreements offered in the Nazareth Land Consolidation project in Belgium, to thousands (over 8,000 in the English ESAs and over 50,000 in the MEKA scheme in Germany). Average figures for annual administrative costs per hectare, per scheme-participant and per ECU paid as compensation were calculated for all of the case-study schemes in each member state across the period of study for each scheme, to summarise the time series data collected in the member state case studies; see Table 4.3. The annual administrative costs for each scheme were weighted according to its value for each given denominator (for example, hectares entered), rather than simply taking the arithmetic average, in order to take into account scheme scale.

Although data were very limited, agri-environmental schemes appeared generally to be more costly to administer, relative to other types of policy such as the commodity regimes for farm income support. It is relatively easy to transfer funds to land-owners, but much more difficult to ensure that environmental conditions are followed in return. Fragmentary evidence of the costs of implementing agricultural commodity regimes in the UK, Germany and Sweden is given in Table 4.4. Agri-environmental schemes generally involve more direct interaction with farmers at all stages than other policy types. There is often a need for substantial professional input from

Table 4.4
The transactions costs of agricultural commodity regimes

		Administration as a % of total public scheme costs
Germany (1993)	Arable area payments	4
	Livestock	20
UK (1996)	Arable area payments	0.8
	Set-aside	3.4
	All crops and set-aside	1.4
	Beef payments	4.9
	Sheep	2.5
Sweden (1997)	Arable area payments	3
	Livestock payments	4

Sources: Lampe (1994); Kumm and Drake (1998); MAFF/IBAP (1997); (see also NAO, 1999).

NB: These costs are calculated on a slightly different basis from those in Table 4.3 in that transactions costs are expressed as a per cent of total, not compensation costs. The data for Germany relate to the first year of the scheme; costs were thought to fall in subsequent years.

project officers (in terms of promoting the scheme, negotiating management agreements, helping farmers adjust their practices as required and so on) to ensure agri-environmental scheme success, given the sometimes complex changes in management needed. Some agri-environmental schemes, such as KULAP in Germany and *prime à l'herbe* in France, implicitly have a strong income orientation, and are characterised by broad approaches covering the whole territory. They were observed to have low administrative costs relative to total scheme spending compared to other agri-environmental policies. Defenders of such schemes have argued that they do at least raise general awareness amongst farmers of the need to manage the environment more carefully, and so offer higher long-run value for money.

Average annual administrative costs per hectare ranged from 9–76 ECU. Untangling the reasons for such variation is complex. Agri-environmental schemes in Belgium have been implemented on a relatively small scale, for example, compared to Sweden, which may account for some of the cost difference given the existence of substantial fixed costs and the relative impacts of economies of scale. Schemes in Italy were observed to have very low estimated costs, which may reflect the early stage of development of the agri-environmental policy framework and the fairly simple administrative structure in place.

Average annual administrative costs per participant ranged from 140–2446 ECU, with average levels typically of 200–300 ECU per participant. The UK schemes had very high estimated costs per participant, perhaps reflecting the number of schemes in place and thus the large proportion of overall administrative spending in the agri-environmental sphere covering administrative overheads. The large average size of farm in the UK may also contribute to this result. The smallest schemes (which have experienced low take-up to date) have now been merged, which should allow a substantial fall in average annual per-unit costs (see LUC, 1995). The UK agri-environmental agencies also spend a considerable amount on environmental monitoring, the costs of which activity were generally not measured for the other country case-studies; as a result, the cost estimates for the UK were inflated relative to costs elsewhere.

Administrative expenditures as a proportion of scheme compensation costs varied from 6–87%. The average for Austria, Germany, Greece, Italy and Sweden was around 10% of compensation costs, with a much higher figure for Belgium, France and the UK. The Belgian data-set included information for a scheme for which no compensation payments were made, thus raising the average proportion on administration. Belgium also has a very small agri-environmental sector over which to spread the fixed administrative costs of policy evaluation and development. France has an extensive and deeply entrenched administrative infrastructure, which may explain its relatively high expenditure. The UK data-set included figures for some very small-scale but high-overhead schemes.

An hypothesis linked to the critical fixed cost/variable cost distinction in overall scheme administrative costs was that per-unit administrative costs should fall with time over the period following scheme implementation. New schemes typically require fixed-cost set-up administrative activities in their first year, rather than the transactional activities relating to land-owner participation. Transactional activities rise in relative importance once participation increases. The marginal increase in scheme participation generally falls with time, and thus, a few years after implementation, most administrative work will relate to maintenance of existing management agreements, rather than the more resource-intensive work of setting-up new agreements. Different contracting and administrative activities are needed at different stages, so there is an uneven time profile of costs. After the set-up period, payments to participants begin to flow, and although monitoring and running costs become quite significant in absolute terms, the relative importance of administrative costs dwindles. In addition, experience in the years following implementation should allow efficiency to be improved. Growing administrative experience would also be expected to impact upon cost-levels, in terms of fine-tuning and efficiency improvements, perhaps from technological developments such as computerisation. However, it is impossible to untangle the relative contributions of these factors in a rigorous way, given the small amount of data available at present.

Nevertheless, there was a notable downward trend in the observed annual administrative costs per hectare over time. A similar trend was noted for per-participant administrative costs, as would be expected (although the area entered into a scheme and the number of participants was not perfectly correlated). However, time series analysis across time was limited in that few data were available for costs in the fourth year and beyond; most countries could supply data only for the first three years of schemes, at most. A downward trend in the staff-hours required per hectare entered into any given scheme over time was also noted, as shown in fig. 4.2. Typically, most schemes had an annual time usage level of less than 20 hours per hectare, and less than 200 hours per agreement made. In France, the arable conversion scheme used around 0.2 hours per hectare, compared to 75–140 hours per hectare for the long-term set-aside scheme, which is implemented on a very small scale. In the UK, most schemes took around 1–3 hours per hectare, with greater inputs demanded by the Countryside Access Scheme (7–10 hours). In Austria, the Ecologically Valuable Areas scheme took around 2 hours per agreement annually, compared to 10 hours for the organic aid scheme, and 50–60 for the eco-points scheme. Scheme characteristics are likely to affect the time requirements, although the sample of schemes was too small to be able to assess this on a systematic basis.

The relative importance of administrative costs to overall scheme costs funded from the public exchequer varies from a very low percentage (1–3%) to 100% of scheme expenditure at the limit.[4] Some schemes had very high ratios due to their low participation levels (such as the Moorland and Habitat schemes in the UK). A large proportion of the annual administrative costs of such schemes are fixed costs relating to policy monitoring and evaluation. Generally, percentages (the relative importance of administrative costs to overall scheme expenditure) were observed to fall with time following scheme implementation, from a range of 1–90% to around 1–50% (with levels typically of 10–20%), after the first few years of the scheme.

---

[4] The Belgian *Bocage Ardennais* scheme was the only case study for which no compensatory payments are made.

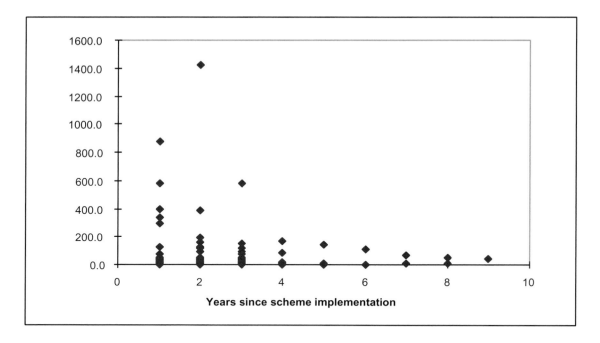

Fig. 4.2. **Annual administrative time requirements (staff hours) per participant in schemes, over time.**

The ratio of scheme administration costs to the compensation payments in each year is of some interest, as it indicates by how much typically the public exchequer costs of schemes are being under-reported, for example, in government policy evaluation documents. All scheme costs, whether compensatory payments or administrative costs, are borne by the tax-payer and therefore represent social opportunity costs in terms of the options to achieve welfare improvements through other forms of policy expenditure[5]. Significant variability in this parameter was observed across the case-study schemes, and across the time span of each case study. This ratio also falls with time; some schemes had extremely high ratios in their early years. For example, the UK Countryside Access scheme, which experienced much lower participation levels than expected initially, had a ratio of 350% in 1996/7.

The observed relationship between scheme participation, in terms of the cumulative hectares entered into a scheme, and the cumulative number of participants (numbers of management agreements), with per-unit administration costs gave some support to a hypothesis that economies of scale exist in agri-environmental scheme administration, as expected given the substantial fixed costs of scheme overheads (see fig. 4.3). For example, in Germany, the MEKA scheme had the lowest administrative costs, relative to total public expenditure on a scheme, and in terms of absolute expenditure; it has a very high acceptance rate with a large area under the policy and high compensatory payments. The FUL scheme (Part II measures) was the most expensive scheme to operate in Germany; its high costs were attributed to low participation (a small area is covered by the scheme) and the fact that payments are targeted on selected plots.

---

[5] In some cases, though, the social opportunity costs may be negative, where a scheme is reducing a distortion created by another policy (see Saunders, 1996).

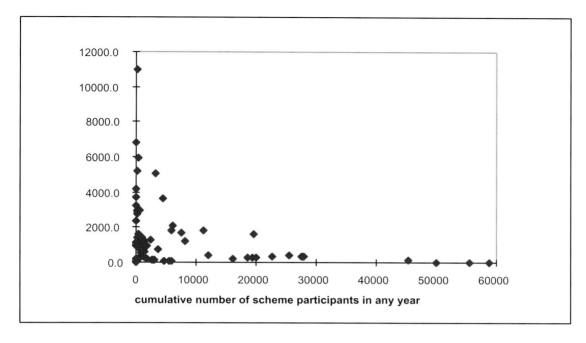

Fig. 4.3. **Annual administrative costs (ECU) per participant and scheme scale (cumulative participation levels) in each year**[6]

However, it is very difficult with existing aggregated data to separate out the scale-economy factor from other factors such as growth in experience in running schemes and fine-tuning. For example, MEKA has a relatively simple scheme application procedure, which was further streamlined in 1993 when the administration of payments under it was integrated with the agricultural support payment system. At present, there is simply insufficient data on scheme organisation costs following the set-up stage for further investigation into scale economies to be possible. In principle, it is possible that some diseconomies of scale will appear as schemes develop, although no evidence of this was found in the present data-set. In addition, over time, fewer agreements are generally made (entry levels are of course finite), and it is very likely that negotiations become increasingly complex, as the more 'straightforward' agreements would be expected to have been made much earlier (i.e. agreements with lower agricultural opportunity costs or ones with lower transactions costs). Although there might be some scope for economies given the growth in the experience of administrators of negotiating management agreements, cost reductions may be small as the site-specificity of individual farms and the choice faced by farmers of different sets of prescriptions mean that blue-prints for the set-up and monitoring of agreements can only go so far. Thus, we might expect the marginal administrative costs of new scheme participants to rise over time, opposing the cost-reducing impact of any economies of scale. This subject deserves more investigation in the future, as scheme participation continues to rise.

In developing programmes under Regulation 2078/92, member states were allowed a significant amount of flexibility to reflect the diversity of their environmental conditions, priorities, agricultural structures and so on. Even schemes of the same name (such as organic farming aid) are far from uniform in their structure. Restricting policy analysis to comparison of the estimated

---

[6] Excluding the UK moorland scheme, for which costs were 44,000 ECU per participant in its second year.

Table 4.5
Organic aid schemes compared, by member states and by years since implementation

	Year	Administrative costs as a % of total costs in each year	Hectares	Number of agreements	Annual administrative costs, ECU per hectare	Annual administrative costs, ECU per agreement
Austria	1995	7.37	259,588	18,543	1.3	18.2
	1996	7.35	272,062	19,433	1.3	18.6
	1997	7.33	280,000	20,000	1.4	19.0
Belgium	1994	3.68	2219	95	8.3	194.8
	1995	3.58	2690	109	8.9	219.5
	1996	2.39	3591	143	5.8	146.3
England	1994	91.13	0	0		
	1995	42.55	4673	101	51.6	2389.3
	1996	30.21	7875	170	30.7	1424.0
France	1994	28.35	18,850	732	51.9	1337.7
	1995	27.61	20,324	783	50.1	1300.3
	1996	29.79	32,331	1417	55.9	1275.4
Greece	1997	10.60	4000	837	119.8	572.5

administrative costs of similar types of schemes has its advantages, namely of increasing the numbers of degrees of freedom available, which is important given a small data-set. It is useful to focus on a specific scheme type, given the heterogeneity of the case studies. Organic aid schemes were provided as case studies for five of the eight member states. However, although the schemes have the same structure and objectives, wide variability in administrative resource use was still observed (as shown in Table 4.5, from the first year of the scheme in each case). Various factors must be considered in addition to scheme type: for example, the stage of implementation of the scheme; the potential for economies of scale; the rigour of compliance monitoring procedures; the degree to which some transactional costs have been passed on to scheme participants through certification and reporting requirements, and so on. Further investigation of the finer details of the schemes is therefore needed.

4.5. Discussion

4.5.1. General factors affecting administrative cost levels

This chapter has focused on the administrative costs incurred in relation to policy implementation to correct for agri-environmental externalities, i.e. on the broad sub-set of transactions costs containing those costs incurred in interactions between the State and scheme participants, rather than between different agencies of the State or individuals in private markets. It appears that agri-environmental payment schemes are more costly to administer than agricultural income supports; in addition there is great variability in the administrative resource use of different schemes. Understanding these observations could help rationalise scheme design and save resources. In addition, setting the policy-related transactional costs against scheme effectiveness in providing agri-environmental benefits and achieving their objectives later, could improve policy evaluation and ultimately improve the value for money of public expenditure in the agri-environmental sphere. Although the agri-environmental policy sphere is relatively small at present, it looks set to expand in the future, so it is important to think through its resource implications, including organisational aspects. Specific attention should be given to the attributes of different transactions and their implications for economic organisation; this is an area for future work. Some more general factors affecting administrative cost levels are summarised below:

- scheme transparency and the ease with which environmental management requirements are understood by farmers, without needing recourse to expert, professional advice;
- the observability of compliance with the environmental management requirements (linked to the nature of the agri-environmental goods to be provided). Further work to link these attributes with the criteria developed in the previous chapter would be valuable;
- scheme objectives and the degree to which these are pursued (i.e. the difference between simply giving farmers compensatory payments, and ensuring that they actually change their management practices to generate environmental benefits);
- the degree of targeting and site-specific negotiations (given the likely trade-offs between compensation and organisational costs);
- the regularity of interactions between regulators and participants, for example, in terms of the frequency of compliance monitoring;
- the potential for economies of scale, given substantial fixed costs of scheme set-up;
- the time since scheme implementation, linked both to the activities required for scheme administration and the likelihood of fine-tuning and efficiency improvements from the experience gained by the implementing agency in the earlier stages;
- the technology available for monitoring and administration (for example, geographic information systems linked to databases to avoid any duplication of payments under different but related schemes);
- farmer attitudes and understanding, which may affect the level of monitoring and enforcement activities needed, insofar as better understanding by farmers would lead to fewer breaches of contract.

Different member states have different concerns, in terms of both environmental conservation and agricultural production support, and these differences might be expected to have implications for policy transactions costs. The traditions of policy-making and styles of regulation differ, and the integration of environmental objectives into the general policy framework takes time, proceeding at different rates in different member states, sectors and so on. Environmental ministries have only recently been established in most member states; the oldest have been in existence for less than 25 years, and in some (as in Italy and Spain), for less than 10 (Wilkinson, 1997).

The context, and stage, of policy development is undoubtedly important to administrative structures and efficiency. We might expect administrative costs to be relatively less important for programmes which were in place some years previously compared to schemes that have been introduced more recently (fewer set-up activities needed, experience of running agri-environmental schemes and so on). However, there is a difference between 'lean' administration and 'streamlined' administration resulting from experience and fine-tuning, so caution in analysis is needed. For example, administration costs in Greece appeared to be low relative to those in the UK, which may be due to the high investment in the agri-environmental policy infrastructure (notably the environmental monitoring strategy) over the past decade in the latter; costs may rise with policy development, rather than fall. It is critical, therefore, to assess the marginal productivity of extra units of expenditure on organisation.

4.5.2. Where does the burden of transactional costs fall?

An important issue in the context of assessing organisational resource use relates to the EU dimension in agricultural and agri-environmental policy development. In addition to the absolute magnitude of administrative costs as items of government expenditure, some of their economic importance stems from the fact that policy administration is the only element of costs that must be borne entirely at the member-state level, even when the policy is to fulfil EU requirements (notably under Regulations 2078/92 and 946/96). This dimension gives rise to two main considerations.

First, some member-states may be constrained financially in developing their agri-environmental frameworks, and particularly in terms of their implementation of European regulations and directives. One implication is that some of the administrative costs of agri-environmental schemes should, perhaps, be covered by transfers from the EU budget, if EU objectives are to be achieved. The treatment of transactions costs by the EU for Regulation 2078/92 contrasts with the financial aid distributed to assist compliance by government agencies with the Natura 2000 network of protected sites throughout the EU. The Natura 2000 implementation costs, which differ from expenditures under the Agri-Environment Regulation in that they contain substantial capital works undertaken in national and natural parks and entail relatively little compensation to farmers and land-users, are to be re-imbursed to the relevant member states under the LIFE instrument (Regulation 1973/92). The provision could perhaps provide a precedent for the greater coverage by EU co-funding of scheme organisational costs in future. In any case, it is evident that the EU is already picking up some transactions costs through the provisions to reimburse member states partly for the compensation payments made to farmers, to the extent that these costs also cover private transactions costs, as discussed above.

Second, given the current institutional framework and provisions for co-funding, policy transactions costs may have an important economic implication through the creation of an incentive for member-states to favour, to some degree, low transactions-cost but high compensation-cost policies (as scheme compensation costs may be reimbursed at a 50% or 75% rate from the EU, under Regulation 2078/92) where possible. Policy frameworks that are potentially inefficient in terms of overall value for money may result, as member states target agri-environmental policies and differentiate payments (according to natural heritage production and/or agricultural opportunity costs) to a lower degree.

4.5.3. What scope is there to reduce the administrative costs of agri-environmental schemes?

The underlying question is, how agri-environmental policies can be designed to minimise public administration and private transactions costs, in relation to delivery of benefits, i.e. with regard to the optimal allocation of public resources to schemes and the division of expenditure between administrative costs and participant compensation payments. Is it possible to improve upon the status quo? Administrative resource use does not necessarily imply inefficiency, even when high levels of costs are incurred (in absolute or in relative terms). The importance of some administrative activities, if not all of them, to scheme success has been emphasised here: it is crucial to keep such costs in perspective, while assessing whether available funding is being utilised effectively. Much more detail of individual schemes and administering units is needed for precise recommendations to be made for practical administrative efficiency improvements. However, there are a number of general options which may allow administrative economies to be made, particularly given the likely increase in the importance of administrative costs to the member state Exchequers if reliance on the use of management-agreement approaches increases in the future:

- reduce the frequency of participant monitoring through the introduction of significant penalties for non-compliance (see Russell 1990)
- increase scheme promotional activities, so participants have a better understanding of the management requirements; which may result in a reduced need for compliance activities. Increased promotion and understanding may also increase interest in scheme participation, allowing some economies of scale
- contract out some routine administrative activities through competitive tendering to reduce costs
- build on existing institutions when developing policies to reduce costs in absolute terms, given the largely fixed costs of developing scheme implementation frameworks, and develop a system of 'one-stop' shops for agri-environmental schemes
- develop the joint administration of agricultural support and agri-environmental schemes (particularly using the IACs (Integrated Administration and Control) system, as in Austria and in some German Länder)

The incremental and cumulative numbers of agreements are important determinants of costs, given the link of some cost-components to participation levels and changes in these, and the fact that management agreements require two-way interactions between landholders and the administering agency. It appears to be far more expensive to establish agreements than to maintain them; the set-up costs should be considered as an investment, and the life-span of the 'asset' (the natural capital protected and maintained or enhanced, and perhaps information relating to conservation activities) should be considered in relation to these costs.

Agreements should be maintained for a period long enough to gain real benefits from their set-up costs. There is substantial investment in human capital as a result of the transacting process: loss of such capital should be avoided where possible; the continuity of the relationship between farmers and project officers may be very valuable and a long-term perspective should be encouraged for both project officers and participants. Agencies should also try to build on successful management agreements incrementally, perhaps under an umbrella agri-environmental scheme rather than starting negotiations afresh for individual schemes, or through the development of a single agency to administer all agri-environmental schemes (one-stop shop). Movement towards such a system has been observed, for example, in Wales and Scotland.

The dynamic nature of the agri-environmental policy context must be borne in mind: in particular, it is hypothesised that the mere existence of schemes may help to bring farming and conservation objectives more in line over time, hence scheme establishment and running should become easier as consensus grows regarding the most appropriate means and ends of policy. Furthermore, the learning curve and fine-tuning in administrative procedures and broader scheme design cannot be ignored in the context of transactional-cost analysis. Substantial development in human capital in the early stages may help transactions costs to fall over time. Institution-building is central (in both the public and the private sectors). The role of experience and education in the production of natural capital, at farm-level and through policy mechanisms, should not be under-estimated, particularly if co-operative approaches rather than regulatory strategies are to develop to increase the levels of agri-environmental goods provision.

The re-design of agri-environmental policy mechanisms should also be considered, from a transactional-cost perspective. For example, a greater differentiation of payment rates could be introduced, to take account of the varying compliance costs of scheme participants, and to tie payments more closely to environmental performance, perhaps using some form of competitive bidding. The problem is that the introduction of more targeted schemes, with management prescriptions tailored more precisely to the conditions of individual sites, will undoubtedly increase administrative costs, although with the aim of increasing environmental benefits too. It may be the case that public administrative costs and participant compensation payments are to some degree substitutes for each other, so an area for further research relates to their relationship, and how direct and how linear any trade-offs are. Hence there is an utility of knowledge of the costs of simpler mechanisms, such as uniform payments. There may or may not be a pay-off overall, taking both production and organisational costs into account, from increasing the targeting of payments under agri-environmental schemes. Further work is needed to investigate this concern; the STEWPOL work has provided some relevant data.

Furthermore, given some degree of joint production of some agri-environmental goods, such as stonewalls and landscape, co-ordinated action by all land-owners in the area is ideally needed, but so far little has been achieved in this area of policy-making. The development of farmer networks could be promoted, to attempt to achieve conservation improvements from peer-pressure and better information (through sharing knowledge) on conservation land management, before moving to collective management agreements (for example, with a premium to encourage participation) (see MacFarlane and Smith, 1998). Collective management agreements, for example, under the existing ESA model, will in all likelihood entail greater administrative costs than individual management agreements. There is some evidence from data relating to the administration of the English ESAs that extra administrative costs arise from the need to co-ordinate farmers' participation in relation to the development of management agreements over common land in ESAs (Falconer and Whitby, 1999a).

This chapter has focused on the public-sector transactional costs of agri-environmental policy implementation. The private costs of participating in countryside stewardship schemes, and the extent to which these may be constraints on policy participation, need examination. Given significant private transactions costs, it is necessary either to rely on farmer altruism, or to increase compensation payments to take account of these costs, if participation is to be at the levels expected based on foregone agricultural incomes. Given the voluntary basis of many countryside stewardship schemes at present, participation levels are a key to their success, so we need to understand more about what motivates or constrains participation in schemes for which the landowner is eligible (see also Chapter 5).

4.5.4. Scheme value for money

Transactions costs are measures only of inputs to achieve policy goals; some complementary measure of output is needed, and the absence at present of such a measure is a significant constraint on economic analysis. A clear distinction between financial analysis and broader economic analysis is crucial in work such as this on administrative-costs. The underlying issue relates to the identification of the best way to achieve agri-environmental objectives. Hence the real interest lies in the change that a given policy, through its attendant expenditure, can achieve, in terms of the environmental benefits flowing from a given amount of expenditure on compensation and scheme organisation. It is unrewarding to discuss transactions costs in the abstract: ultimately, analysis must be related to the policy objectives and the extent of their achievement if decisions are to be made to maximise economic efficiency in resource allocation.

Optimal policy involves achieving a balance between the marginal costs and the marginal benefits of policy intervention. The essential implications of empirical administrative cost research presented here is that if policy costs are higher than originally thought (or officially presented) because of the costs of policy organisation, we need to ensure that benefits are still higher than the new estimates, if policy intervention is to be cost-effective. However, full cost-benefit assessment is prohibitively complex. Even if we knew exactly what physical changes were occurring in the countryside and the rural environment, where, and when, there are problems in linking cause and effect, given the multitude of different policies and economic forces at work. It is almost impossible to untangle the consequences of any particular scheme with a great level of certainty.

In addition to the methodological problems associated with environmental monitoring, there are conceptual issues in relation to benefit measurement: for example, is the maintenance of the status quo really a benefit to society or not? The answer depends on what would have happened in the absence of the policy and on the perspective taken, using the notion of the 'reference point' (Bromley and Hodge, 1990) (see also Chapter 2). Still, it is important to stress the potential benefits of administrative expenditure in the agri-environmental policy sphere: after all, the intention of government intervention is to increase social welfare. However, for this to be the case, great care is needed to ensure that the administrative activities that take place are the most appropriate ones.

4.6. Conclusions

Some level of administrative transactions costs is both inevitable and necessary if externalities are to be optimised through policy implementation. All forms of agri-environmental policy intervention, irrespective of their form, will cause administrative costs to be incurred, so it is a potentially serious mistake to ignore the costs involved. Understanding policy transacting in more depth, especially through empirical research, is timely given the expansion of the agri-environmental sector, and the possible increases in expenditure upon it. Policy-related transactions costs comprise significant, and neglected, elements of public expenditure, estimated at 1–2 billion ECU across the EU for Regulation 2078/92 schemes since their implementation in 1992 to 1996/7, based on the cost assessments made for case-study schemes across eight member states. This compares with the 6.9 billion ECU paid in compensation to farmers over the same period. These costs are often overlooked, yet they may be sufficiently important to constrain the resources available for implementing such policies, especially in times of public expenditure scrutiny and cut-back. Greater transparency with regard to administrative costs is required as a safeguard against inappropriate public policy spending and lower levels of overall social welfare than might otherwise be possible.

Agri-environmental policies are still in their infancy. Organisational costs may well become of even greater importance in the future as policy objectives and the mechanisms used to achieve them evolve: budgets should be set in the light of information on the full public costs of such policies, particularly in the context of proposed changes in the mix of schemes under Agenda 2000. This will apply particularly if, for example, 'greening' of LFA payments or environmental cross-compliance are to be widely introduced. High transactions costs will influence policy progress; however, it is very difficult, at present, to assess ways of reducing costs, especially as the policy context is so dynamic. There are questions relating not just to the levels of administrative expenditures on agri-environmental schemes, but also to the mix of activities funded by such expenditures. Trade-offs will in all likelihood be required: scope for economising in one area will be balanced by increased requirements in another. It is recognised that it is still early days for scheme evaluation; furthermore, it is essential not to lose sight of the underlying goals of policies, namely environmental improvements. The current agri-environmental policy framework still falls short of a fully-integrated approach to environmental management; further development is likely, and the message of this study is that the administrative cost implications should be considered as an integral part of public policy decision-making.

References

Arrow, K. (1969) The Organisation of Economic Activity: Issues Pertinent to the Choice of Market Versus Non-Market Allocation. In: *The Analysis and Evaluation of Public Expenditure: the PPB System.* Vol. 1, US Joint Economic Committee, 91st Congress, US Government Printing Office, Washington D.C.

Bromley, D. and Hodge, I. (1990) Private Property Rights and Presumptive Entitlements: Reconsidering the Premises of Rural Policy, *European Review of Agricultural Economics*, 17, 197–214.

Cheung, S. (1987) Economic Organisation and Transactions Costs. In: J. Eatwell, M. Milgate and Newman P. (Eds), *Palgrave: New Dictionary of Economics.* MacMillan, Basingstoke.

Coase, R. (1960) The Problem of Social Cost, *Journal of Law and Economics*, 6, 144.

Dahlman, C. (1979) The Problem of Externality, *Journal of Law and Economics*, 22, 141–162.

Falconer, K. and Whitby, M. (1999a) *Administrative Costs in Agricultural Policies: the case of the English Environmentally Sensitive Areas.* Research report of the Centre for Rural Economy, University of Newcastle upon Tyne.

Falconer, K. and Whitby, M. (1999b) *Transactions and Administrative Costs of Countryside Stewardship Policies: an Investigation for Eight European Member States.* Working paper, Centre for Rural Economy, University of Newcastle upon Tyne.

Kumm, K. I. and Drake, L. (1998) *Transaction Costs To Farmers Of Environmental Compensation.* Department of Economics, SLU, Uppsala, Sweden, (unpublished paper).

Lampe, K. (1994) *Mögliche Auswirkungen der EU-Agrarreform auf die Agrastruktur und die Entwicklung des Ländlichen Raumes in Niedersächsischen Teil der Eurigio.* Eurigio-Symposium, Bad Bentheim.

Latacz-Lohman, U. and Van der Hamsvoort, C. P. C. M. (1998) Auctions as a Means of Creating a Market for Public Goods for Agriculture. *Journal of Agricultural Economics*, 49(3), 334–345.

LUC (Land Use Consultants) (1995) *Countryside Stewardship — Monitoring and Evaluation Third Interim Report to the Countryside Commission.* Countryside Commission, Cheltenham.

MAFF/IBAP (1997) *The Government's Expenditure Plans, 1996/7–1998/9.* CM 1903 HMSO, London.

MacFarlane, R. and Smith, S. (1998) Implementing agri-environmental policy: a landscape ecology perspective. In: A. Cooper and J. Power (Eds), *Species Dispersal and Land-use Processes.* IALE (UK), University of Ulster, Coleraine.

McCann, W. K. and Easter, L. (1998) *Estimating the Transactions Costs of Alternative Policies to Reduce Phosphorous Pollution in the Minnesota River.* Staff paper P98-7, Department of Applied Economics, University of Minnesota.

Moxey, A. and White, B. (1998) Contracts for Regulating Environmental Damage from Farming: a Principal-Agent Approach, *Etudes et Recherches sur les Systèmes Agraires et le Développement*, 31, 205–219.

Moxey, A., White, B. and Ozanne, A. (1999) Efficient Contract design for Agri-environmental Policy, *Journal of Agricultural Economics*, 49.

NAO (National Audit Office) (1997) *Protecting Environmentally Sensitive Areas.* Stationary Office, London.

NAO (National Audit Office) (1999) *Arable Area Payments.* Stationary Office, London.

Niehans, J. (1971) Money and Barter in General Equilibrium with Transactions Costs. *American Economic Review*, 61(5) 773–778.

Russell, C. S. (1990) Game Models for Structuring Monitoring and Enforcement Systems. *Natural Resources Journal*, 4(2), 143–173.

Saunders, C. (1996) *Financial, Public Exchequer and Social Values of Changes in Agricultural Output.* Centre for Rural Economy Working Paper, University of Newcastle upon Tyne.

Stavins, R. (1993) *Transactions Costs and the Performance of Markets for Pollution Control.* Discussion paper QE93-16. Quality of the Environment Division, Resources for the Future, Washington, D.C.

Whitby, M. C. and Saunders, C. M. (1996) Estimating the Supply of Conservation Goods in Britain: a comparison of the financial efficiency of two policy instruments, *Land Economics*, 72(3), 313–325.

Wilkinson, D. (1997) Towards Sustainability in the EU? Steps within the EC Towards Integrating the Environmental into Other EU Policy Sectors. *Environmental Politics*, 6, 153–173.

Williamson, O. E. (1985) *The economic institutions of capitalism: firms, markets, relational contracting.* Free press, New York.

G. Van Huylenbroeck and M. Whitby
Countryside Stewardship: Farmers, Policies and Markets
© 1999 Elsevier Science Ltd. All rights reserved

Chapter 5

Farmers' attitudes and uptake

*Lars Drake, Per Bergström and Henrik Svedsäter**

Abstract—The Chapter reports results of a survey of attitudes to and uptake of CSPs by 2000 farmers in STEWPOL countries. A model of farm household behaviour is developed and the survey methodology explained. Attitude measurement is achieved by inviting respondents to express their acceptance or otherwise of a set of statements about the agri-environment. The results are then aggregated to a Likert score which is examined, using a regression model, explaining attitudes, while participation is analysed using a logit model in terms of perception of and attitudes towards public issues, the schemes available and the structure of the farm business and household. The analysis presents insights into the variation in attitudes towards stewardship policies and hypotheses to explain scheme participation. The survey results underline farmers' generally positive attitudes towards the schemes and the chapter concludes with an examination of possible ways of increasing participation.

5.1. Introduction

The success of any voluntary agri-environmental scheme depends on the level of uptake by farmers. If it is not possible to convince farmers about the benefits of participation, positive aspects of a scheme may be limited in scale. Uptake can be seen as a necessary but not sufficient condition for success. For this reason farmers views and understanding are important in order to improve the functioning of existing policies. The primary objective of this study, labelled "socio-economic analysis of stewardship policies", is to trace out the causes of participation and non-participation in the agri-environmental programme with the purpose of suggesting measures that would increase participation. For this purpose, surveys have been distributed in eight EU countries, focusing on attitudes towards and acceptance of the EU agri-environmental programme (AEP) among farmers.

* We are indebted to Ulf Olsson for statistical guidance.

5.2. Theoretical issues

5.2.1. Model of farm behaviour

As a theoretical framework, a modification of the standard household utility function is proposed in order to illustrate farmers' behaviour with respect to the conservation schemes (standard version from Singh *et al.*, 1986 and Lexmon and Andersson, 1998). In the utility function below it is assumed that farmers pay attention not only to economic incentives, directly or indirectly, when deciding whether or not to enter an agreement, but that in reality, a variety of non-economic factors will also contribute as determinants of participation. The farm household is treated here as one unit.

People can maximise their utility in two different ways, i.e. there are two prerequisites for a rational behaviour. The first condition is that the agent, in this case the farmer, thinks in terms of maximising utility whatever the objectives or the circumstances might be. In neo-classical economic theory it is assumed that this is an inherent objective among humans. For instance, some economists claim that economic factors are the superior determinants in the formation of all sorts of intentions, no matter whether career opportunities or marriage planning are considered (Becker, 1973). However, as previously mentioned, there are other factors entering the equation. Irrespective of our doubt in comparable predictions, we may still argue that the fundamental aim is to maximise utility. The farmer's maximisation problem is stated in Equation 1.

Equation 1: The farmer's maximisation problem

max $u = u(\pi_h, l_l, O_h)$
such that:
$l_l + l_o + l_f + l_e \leq (T-R)$

π_h is the farm household profit, see function 2,
l_l is the amount of leisure in hours/day,
O_h is other positive values derived from the activities within the household profit function,
l_o is the time spent on off farm work in hours/day,
l_f is the time spent on food and fibre (F&F) production in hours/day,
l_e is the time spent on environmental conservation in hours/day, and
$T-R$ is 24 hours (T) minus the number of hours per day needed for physical relaxation.

O_h is a function of all other values derived from the actions within the household profit function. The sense of satisfaction of providing environmental goods may be one source of value included in O_h.

Both in Singh *et al.* (1986) and Lexmon and Andersson (1998), it is shown that

> "the production decisions can be made independently of consumption and leisure decisions. Hence, the marginal value product of the variable input is equal to the factor price. Furthermore, the marginal value product of all labour in the farm operation equals the market wage rate, W_o, in alternative employment, i.e. off-farm labour. Hence, the optimal choice of these production factors is independent of the farmer's preferences for leisure and consumption and the available time allotment T−R"
>
> (Lexmon and Andersson, 1998, p. 31).

Some, probably most, of the environmental conservation efforts are made by changing the technology in the food and fibre production, i.e. environmental effects are generated jointly with F&F. Therefore the separation of time in l_f and l_e is only of theoretical interest. Nevertheless we think that the model may illustrate this on a conceptual level.

The second condition involved has to do with the link between stated intentions and actual behaviour. Although agents in their mind try to and really would like to optimise their behaviour, they may still fail to do this, or simply not have the sufficient ambition to fulfil the given objectives. If this is the case, a cognitive dissonance arises between stated intentions and actual behaviour (Newcomb, 1953). For the ambitious individual this might cause a problem, but we assume that not all human beings perceive the discrepancy as particularly troublesome. Thus, they might admit that one important goal in everyday life is to maximise one's utility, but they might not put enough effort into realising this goal, neither do they bother so much about a possible failure to do so (Newcomb, 1953). The farm household profit is shown in Equation 2.

Equation 2: The farm household profit function

$$\pi_h = P_f f(X_f, l_f) + w_o l_o + C_{ec}Q(.) - P_x X_f - P_{ec}E(.) - w_f l_f$$

- π_h is the farm household profit,
- P_f is the price vector for the F&F goods produced on the farm
- $f(.)$ is the production function for F&F products,
- w_o is the wage for off farm work,
- C_{ec} is the level of compensation payments for environmental conservation efforts,
- $Q(.)$ is the level of environmental conservation provided,
- P_x is the input price vector for F&F production,
- X_f is the inputs for F&F production,
- P_{ec} is the input price vector for environmental conservation efforts,
- $E(.)$ is a function of conservation efforts, see function 3, and
- $w_f l_f$ is the external labour costs for F&F production

The conservation effort $E(.)$ is assumed to be positively correlated with the level of environmental conservation provided, Q, which could be described as $Q = f(E)$. The conservation efforts could be described as a function of various factors associated with a fulfilment of the scheme. Specifically,

Equation 3: The conservation effort function

$$E = f(m, c, tc_p, u(Q))$$

- m is the level of compensation payments,
- c is the direct costs associated with conservation,
- tc_p is the farmers private transaction cost for participating in conservation schemes, and
- $u(.)$ is the direct utility derived from conservation, which obviously is a function of Q and consequently an indirect function of E.

Naturally, m is expected to be positively correlated to E, whereas c and tc_p are expected to be negatively correlated to E. Costs of participation can be either changes in production costs or separate costs for conservation. Apart from the indirect utility derived from compensation through the supply of Q, the agents also derive utility from the level of Q independent of the amount of compensation, assuming that they appreciate a beautiful and varied landscape. Theoretically, the farmers are expected to maximise their utility according to these equations.

A model such as that described might be considered as far-fetched for our purpose of investigating farmers' attitudes towards and uptake of the schemes. The relationship is presumably much more complicated and not easily reduced to a mathematical equation. However, as previously emphasised, the stipulation serves merely as a theoretical point of departure, and not as an attempt to fit the farmers' behaviour into this in detail. The model only serves as the theoretical basis for formulating hypotheses (in Section 5.3.5).

By investigating the relative importance of various factors not directly related to the schemes, we are in a better position to say if, and to what extent, farmers would undertake conservation practices irrespective of the agri-environmental programme. Furthermore, there is naturally a great interest in investigating the contribution of other factors than economic, rational as well as irrational, both from a theoretical and a practical point of view. By gaining insight into this dimension and thereafter better targeting the schemes with respect to factors identified as important, the prospectus of the agri-environmental programme will presumably increase. For instance, the motives among farmers could be individualistic, collective, symbolic, or a combination of these. In case only individualistic motives matter, this implies that information about environmental and cultural values will not alter their behaviour, whereas compensation and other personally important aspects will play a major role.

Secondly, it is not certain that people behave as optimisers, a hypothesis opposite from that to which neo-classical economic theory subscribes. The essential problem regarding this lies in examining if and to what extent people are in fact acting upon rational premises. Although this cannot be tested analytically in this context, we believe it possible to investigate comparable phenomena qualitatively, partly by matching stated intentions and actual behaviour, partly by means of directly asking people about their reasons behind stated intentions in order to trace out a rational goal function.

5.2.2. Attitude measurement

Apart from investigating the attitudes towards the scheme in a qualitative manner, the intention was also to analyse them quantitatively. There are two possible ways of doing this. The most obvious approach among economists is to establish an econometric model, whereby different aspects of farmer's comprehension of the programme are categorised. Then it is possible to draw conclusions upon which factors determine participation and which do not. Another approach which has been employed is to rely on attitude scaling or attitude measurement. Although there are many different possible ways of measuring the strength of a given attitude, Likert scales, also labelled summating rating (e.g. Robson, 1993), have been chosen for this purpose.

Likert scaling is a systematic scaling technique which is well documented (e.g. Eagly and Chaihen, 1993). It has the added advantage of being relatively easy to develop. Furthermore, items in a Likert scale usually look interesting to respondents, and people often enjoy completing a scale like this. Principally, the scale is developed according to the following steps:

- Items apparently related to or important to an issue are gathered together. These should reflect both negative and positive views of the issue. Rather than presented as questions, the items should be presented as statements. An example is "compensation connected to the scheme is too low".
- In the next step, it is up to the respondent to agree or disagree with this statement. A common response categorisation system is to use five fixed alternative expressions, labelled "strongly agree", "agree", "undecided", "disagree" and "strongly disagree".
- A large number of specified items should be presented, rather than a few more general ones. This is because there are usually many aspects or dimensions to an issue such as the agri-environmental programme. In order to yield meaningful answers one has to give the respondents the opportunity to consider these various dimensions separately thus avoiding difficult weighting.

The last step involves the ranking, differentiation and calculation of the strength of the attitude, which will be discussed further below. There are principally three prerequisites for a valid item in the scale; firstly, it has to be relevant by way of presenting significant information. Secondly, it needs to be formulated in an unambiguous way. Finally, it must discriminate between respondents. For instance, if everyone is agreeing with a particular statement, this does not present any substantial information, and should preferably be excluded from the analysis.

5.3. Methodology and data collection

5.3.1. Pre-test

The questionnaire has been rigorously tested before being implemented as a final survey. On the basis of the group discussions and previous research, combined with our own reflections, a first draft of the questionnaire (pre-test questionnaire) was distributed throughout the whole of Sweden and respondents were randomly identified through the Swedish official register of farm business. Altogether 148 surveys were mailed during November 1996, and non-responding farmers were sent a reminding letter three weeks later.

Altogether 65 farmers replied to the questionnaire after one reminding letter, implying a response frequency of 44%. One important feature of the survey is that the responses to each of the questions are balanced, meaning that, on average, "all" response alternatives are relevant. Some questions produced a skewed distribution of responses and were, therefore, either reformulated or excluded from the final survey.

5.3.2. Questionnaire design

The final questionnaire was separated into six sections. The first section consists of some easily answered questions about gender and the respondent's status on the farm.

In the next section the respondents are presented an introductory letter. This information shortly describes how the farm has been selected to participate in the survey and the purpose of the study. In association with this briefing, the respondents are informed that they will be asked what they believe is good and bad with the current schemes and if they have any suggestions of how to improve the agri-environmental programme. This information is important since this presumably raises farmers' interest in the questionnaire, thereby increasing the response rate. As an obvious part of a survey introduction, the respondents are told that their answers will be treated anonymously and in the strictest confidence.

The third section contained only one question—about opinions toward public issues. It has been indicated that comparable items raise the interest on behalf of the participants. Secondly, according to the funnel principle it is recommended to start with general issues and become more specific as the interview proceeds. This is to gain an insight into farmers' views upon environmental values and their relative importance.

Section four contained eleven questions about the respondent's participation in management agreement schemes. Obviously this is a key section since it yields information whether or not a particular farmer has enrolled in, or has intentions to enrol in, any management scheme. This question was followed by questions trying to establish the reasons for high or low participation. An important task is also to identify those farmers whose land is not eligible to enter any management agreement schemes. This group of farmers is less significant for our purposes since they are still not qualified for current management schemes.

The fifth section is concerned with attitudes towards the agri-environmental programme. In this section, a number of statements express various aspects or views of the programme, and it is up to the respondents to agree or disagree with each statement on the basis of a five graded scale. The

Table 5.1
Statements in the questionnaire to test the attitude of farmers towards agri-environmental schemes

1. "The schemes are positive for nature and landscape"
2. "The intended environmental effects are greatly valued by many people"
3. "The state of the environment in the agricultural landscape is critical"
4. "Farmers are already doing enough to protect the environment"
5. "Public compensation to farmers is a viable alternative to market payments"
6. "The schemes are directed toward the most valuable environments in the agricultural landscape"
7. "The schemes are in conflict with other types of agricultural policies"
8. "The information about the schemes have been adequate and sufficient"
9. "The current policy principles and regulations will to a large extent remain over a longer period"
10. "The schemes can be easily implemented on my farm"
11. "The schemes should be more adapted to local circumstances"
12. "The work and time effort required to carry through the conservation measures are too heavy"
13. "The rules that govern who can join the schemes are fair"
14. "The regulations of the schemes are adequately detailed"
15. "The rules and requirements are easy to understand"
16. "The application procedures are too bureaucratic"
17. "There should be more emphasis on agricultural training and practical demonstrations"
18. "The compensation is sufficient to cover the costs incurred by the farmer"
19. "The sanctions for not carrying out the agreements properly are reasonable"
20. "The schemes give rise to negative effects for farm production"

statements are mainly based on what has been expressed by farmers in direct interviews or in the Swedish media. In addition, some items have been formulated on the basis of what initiated experts and the research team believe is important. The statements are used for calculating a Likert average score, which will be used as an index of the respondents' attitude towards the agri-environmental programmes. In Table 5.1 these statements are listed.

Finally, the last section contained questions about the farmer and his/her business, such as farm type, household income level and age of respondent.

5.3.3. Data collection

Needless to say, one of the most important things is that the data collection is made in a neutral manner, thereby minimising the occurrence of compliance bias and interview effects. There are no general guidelines of how this should be done, but we tried through the structure of the questionnaire to establish an atmosphere during the interview whereby respondents feel free to express their own views and opinions.

Our objective was to conduct 250–400 interviews in each country, in order to yield a reasonable representativeness in each country. Only regions where agri-environmental schemes exist were selected. Regions, which differ in terms of farm type and policies, were sampled in order to reflect the array of situations in STEWPOL countries.

Table 5.2
Overview of the whole data set

Country/Land	Region	Number of respondents	Number of represented farmers	Number of strata
France	5 département	349	66,500	15 (département and age)
	Cantal	30	7,400	3 (age)
	Gers	40	11,600	3 (age)
	Isère	39	10,700	3 (age)
	Loire Atlantique	121	15,200	3 (age)
	Calvados/Manche	119	21,600	3 (age)
Austria	Niederösterreich	250	59,243	6 (farm size)
Germany	3 Laender	191	17,959	6 (Laender and districts)
Bavaria	2 districts	66	5,570	2 (districts)
	Neumarkt	35	3,500	1
	Tirschenreuth	31	2,070	1
Saxony	Entire Saxony	56	8,149	1
Schleswig-Holstein	2 districts	69	4,237	2 (districts)
	Dithmarschen	29	2,584	1
	Ostholstein	40	1,653	1
Belgium	2 regions	246	23,541	2 (regions)
	Flanders	117	15,914	
	Walloon	129	7,627	
Greece	Prefecture of Larissa	250	15,000	2 (particip./non parti.)
Italy	2 regions	232	218,901	2 (regions)
	Trentino Alto Adige	70	24,441	1
	Veneto	162	194,460	1
Sweden	The whole country	224	88,028	1
United Kingdom	2 counties in Northern England	255	8,576	6 (county and farm size)
	Northumberland	58	2,320	3 (farm size)
	Cumbria	197	6,256	3 (farm size)
The whole data set	19 regions	1997	497,748	40

Source: Collaborator in each country.

The sampling procedures differ between the countries, but overall there is an acceptable representativeness. The main reason for differences are caused by variations in access to farm registers in the eight countries. A detailed description of the sampling procedures can be found in the main report of the project (Bergström et al., 1999). The study covers 19 regions in eight European countries.

In most countries the sample was stratified, e.g. with respect to region, age or farm type, in order to guarantee that minor sample categories of farmers were not under-represented. The difference in sampling design is possible to correct by using weights, which has been done below in the multiple regression models explaining attitude.

5.3.4. Regression models

The determinants of a positive attitude towards the EU agri-environmental programme are analysed in a multiple regression analysis. The average Likert score, representing attitude, is chosen as the dependent variable.

A regression analysis with participation/non participation as dependent variable has been employed in order to trace out the causes of compliance and non-compliance of the agri-environmental programme. This kind of regression model, with a binary dependent variable, is somewhat different from an ordinary regression model (Gujarati, 1988). Therefore we need a probability model that has two features: (i) as the independent variable X_i increases the probability that the dependent variable has a certain value, $P_i(Y = 1|X)$, increases but never outside the 0–1 interval and (ii) the relationship between P_i and X_i is nonlinear. The logit model has been selected since it has both these features.

5.3.5. Hypotheses

Some hypotheses are developed to capture relations in Equation 3 and some are based on our own judgement. A first expectation is that the opportunity cost of participating in the EU agri-environmental programme, i.e. reduction of profits from F&F production or increased need for investment or work effort, will have a negative impact on participation, even if this influence is fairly weak (Hypothesis: Economic).

The average Likert scores are expected to be significantly higher among participants than among non-participants. We believe participation is positively affected by high scores on the attitude questions concerning environmental, administrative, economic or general aspects of the programme (Hypothesis: Attitude).

The following group of hypotheses is related to the problem of transaction costs. Private transaction costs for participation are relatively high (15% of compensation payments estimated for Sweden in Kumm and Drake, 1998) and as explained in Chapter 4 there are fixed costs that are independent of size or extent of effort. Therefore an hypothesis is that small farms would sustain relatively higher transaction costs than larger farms (Hypothesis: size). This may be counter-acted if a larger proportion of small farmers view farming as a way of life rather than as an economic enterprise. The hypothesis in relation to this is that large farms are more likely to participate than smaller farms. The probability for participation is also expected to be higher if the farmer has previously been involved in conservation schemes. Those farmers have experience which may make them more positive and reduce their transaction costs. They may also experience less costs since some adaptations have already taken place (Hypothesis: experience). We also believe that the transaction costs will be relatively lower for a farmer with a long general education compared with a farmer with a shorter education. The latter group of farmers may have to use more expert help which increases the private transaction costs (Hypothesis: education). Also, education may increase environmental concern.

The farmers are likely to discuss the agri-environmental programme with relatives and neighbours and one's participation may increase the probability of another to participate (Hypothesis: Neighbour).

Household income may also affect the decision to participate. Low income may result in a greater incentive to participate. However, counter-active factors such as that high income groups tend to be more interested or concerned about environmental issues. Therefore the hypothesis is that income has a positive influence on participation (Hypothesis: Income).

Quite a large fraction of the agri-environmental programme is targeted to animal production and therefore the animal farms are expected to be more involved in the programmes (Hypothesis: Animal).

Age of the respondents turns out to be a significant explaining variable in many surveys. Younger farmers can be expected to be more open towards the agri-environmental programme which represent a new policy approach. Older farmers may have a greater interest in conserving landscape that they have been working in for a long time. Overall our hypothesis is that younger farmers have a higher participation rate than older (Hypothesis: Age).

Our last hypothesis is that there are significant differences among countries due to difference in general farming conditions, schemes (compensation levels etc.) and attitudes (Hypothesis: Country).

5.4. Analysis of results

5.4.1. Some general results

Farmers have in general indicated a positive attitude towards the agri-environmental programme. They also find themselves well informed about the schemes. The two statements on the Likert scale that received the strongest positive response were 'the schemes are positive for nature and landscape' and 'the state of the environment in the agricultural landscape is critical'. The two statements on the Likert scale that received the strongest negative response were 'there should be more emphasis on agricultural training and practical demonstrations' and 'the schemes should be more adapted to local circumstances'. In general there was a positive attitude towards the EU agri-environmental programme.

Four questions addressed the respondents' expectations of the effect on farm business of the agri-environmental programme. The issue is whether production, profit, hours worked or investment would increase, decrease or remain unchanged. In all four cases the highest frequency is indicated for the status quo alternative. For respondents who did indicate a change, the tendency is that production is expected to be reduced while profits, hours worked and investment will increase. Hours worked would include both changes in production and transaction costs. The difference between frequency for increase and decrease is as follows: production -0.20; profit $+0.03$; hours worked $+0.34$; and investment $+0.25$.

For the whole data set, which is labelled 'EU8', medical service, school systems and environmental preservation are the three (out of eight) most important public issues. There are, however, important country differences on this topic. In Belgium, Germany and United Kingdom school system is ranked the highest, in Greece and Sweden medical service, in France unemployment and in Italy environmental preservation received the highest ranking. Respondents in Austria rank environmental preservation as high as social security and pensions.

Reasons for participation in the agri-environmental programme tend to be similar in the different countries and regions (Table 5.3). When possible, the data for Germany have been divided into the three selected Länder since the conditions, including agri-environmental programmes, vary among them. Environmental concern is ranked highest in the EU8 and in eight out of ten regions, which indicates a support for Hypothesis: Attitude. In seven out of ten regions, economic incentives were ranked second or first. The third most important motive, i.e. that the schemes fitted in with future management plans, is also an economic motive since it reduces the costs of participation. These results indicate a support for Hypothesis: Economic. The other motives have received lower rankings in practically all regions. For some countries, participation in previous schemes was ranked relatively high.

Another question which is related to the motive that the schemes fitted in with future management plans, is whether respondents would 'continue undertaking the conservation measures in the schemes regardless of the future level of compensation'. Forty-one per cent would do this to a large extent and 44% to a limited extent. This raises the question of how far farmers are compensated for efforts they are already doing rather than for new efforts. Even if there is some over-compensation in the short run, the subsidies may be needed in the long run in order to attain the goals (Kumm and Drake, 1998).

The main reason for not participating is, as shown in Table 5.4, that respondents lack information. The second most important motive is that the farmer believes compensation rates are too low, which supports Hypothesis: Economic. In Sweden, application procedure is thought to be too complicated. Saxony in Germany and Austria were not included in this analysis due to the small number of non-participants in the survey. Respondents whose application was refused were excluded from this analysis since they have not chosen to stay out of the programme.

As shown in Table 5.5, the hypothesis that attitudes of participants are more positive than those of non-participants (Hypothesis: Attitude) is accepted for EU8 and for all countries except for Sweden and Austria. The hypothesis for UK is accepted when considering the second decimal place. Austria is not analysed as a single country, since the number of non-participants among the respondents are only 5 which is far too small to make a statistically reliable test. It is, thus, shown that in six out of seven analysed countries participants are more positive towards the agri-environmental programme than non-participants. The positive relationship between participation and attitude towards the programme shows that attitude as measured by the Likert scale captures attitude variation in an expected way. It is thus reasonable to use it as a general indicator of attitude towards the agri-environmental programme.

5.4.2. Model explaining attitude

The dependent variable for all models explaining attitude is the average score on the Likert scale. The responses have been weighted due to stratified sampling procedures. The weighting procedure is described in detail in Bergström *et al.* (1999). The search for the best models started with a complete model containing all relevant variables. From this model the variable which suites the model the least due to its absolute t-value has been excluded. In order to ensure that the new model is not under-specified, the change in F-value has been checked. This process of excluding variables is stopped when the absolute t-value of the least contributing variable is higher than 1.4 which gives a maximum P-value of approximately 0.150.

The first model of the whole data set, presented in Table 5.6, uses 1654 responses while 343 responses are rejected due to missing values for at least one variable. This model includes seven country dummies (eight countries minus Belgium, the omitted reference). All seven country dummies receive a positive sign for their β-value. This is explained by a substantial decrease of the value of the intercept when the country dummies enter the model. The French-dummy

Table 5.3
Reasons for participation in management agreement schemes, ranking*

Geographical area	Participants who have indicated:							
	Participated in previous scheme	Scheme fits in with future management plans	Maintain or improve the natural environment	Economic incentive offered by scheme(s)	Encouragement by friend or neighbour	Encouragement by officials	Scaling down farming activities anyway	Other reason
			As a reason to participate					
France	5	3	1	2	6	6	8	4
Austria	4	3	1	2	8	5	6	6
Bavaria**	8	2	1	3	5	8	8	4
Saxony**	5	3	1	2	7	5	8	4
Schleswig-Holstein**	8	2	3	1	4	8	8	8
Belgium	5	2	1	3	8	4	7	6
Greece	5	4	2	1	7	6	8	3
Italy	5	4	1	2	7	3	8	6
Sweden	4	2	1	3	8	5	5	5
United Kingdom	8	2	1	2	7	5	5	4
EU8	4	3	1	2	7	5	7	6
Average score	0.11	0.25	0.47	0.43	0.02	0.09	0.02	0.07

* If the frequency is equal to 0 the alternative will automatically be ranked last (8).
** Regions in Germany.

Table 5.4
Reasons for non-participation in management agreement schemes, rankings*

Geographical area	Non participating respondents who have indicated							
	Compensation rates are too low	Doesn't have the machinery needed for participation	Application procedures are too complicated or costly	Schemes will have a negative impact on my farm output	Practices are too costly in terms of the work requirement	Didn't know enough about the schemes at the time of the application	Concerned that the rules and regulation might change over time	Other reasons
	As one of two main reasons not to participate:							
France	2	8	4	1	7	3	9	12
Austria	–	–	–	–	–	–	–	–
Bavaria**	1	11	11	1	11	11	11	4
Saxony**	–	–	–	–	–	–	–	–
Schleswig-Holstein**	1	11	11	2	11	4	4	6
Belgium	2	7	4	1	6	3	9	11
Greece	1	6	9	3	10	2	7	5
Italy	2	3	4	6	5	1	6	6
Sweden	3	5	2	5	4	5	8	1
United Kingdom	2	11	3	4	9	6	5	1
EU8	2	6	4	5	7	1	8	3
Average score	0.28	0.14	0.19	0.16	0.10	0.44	0.08	0.23

* If the frequency is equal to 0 the alternative will automatically be ranked last (11). The table only includes the eight most important reasons not to participate.
** Regions in Germany

Table 5.5
Difference in attitude (Likert average score) among participants and non-participants

Jurisdictional area	Participates or not	Number of answers	Average Likert score	95% confidence interval
France	Participates	243	0.67	0.01
	Does not participate	106	0.64	0.01
Austria	Participates	245	0.51	0.01
	Does not participate	5	0.58	0.12
Germany	Participates	128	0.53	0.02
	Does not participate	62	0.45	0.03
Belgium	Participates	54	0.44	0.03
	Does not participate	192	0.37	0.01
Greece	Participates	125	0.50	0.01
	Does not participate	125	0.46	0.01
Italy	Participates	115	0.58	0.02
	Does not participate	117	0.52	0.01
Sweden	Participates	138	0.43	0.03
	Does not participate	72	0.43	0.04
United Kingdom	Participates	123	0.56	0.03
	Does not participate	131	0.52	0.01
EU8	Participates	1187	0.54	0.01
	Does not participate	810	0.48	0.01

Source: own calculations.

is very high, i.e. farmers in France express the most positive attitude, and the Swedish dummy is rather low, compared with the others (except Belgium which would have the lowest value had it been included).

Those who rank environmental preservation as the most important public issue tend to have higher scores on the Likert scale. Those who have a participating neighbour or relative tend to be more positive towards the agri-environmental programme. Respondents who have more years of general education and those whose profits have increased during the previous five years tend to have a more positive attitude than others while those with children under the age of 18 tend to be negative.

It is interesting to note that attitude towards agri-environmental programme can be linked to participation in three types of policies. It is not possible to say whether attitude affects participation or the opposite. A good guess is that the cause and effect relation is two-sided. Three statements, not included in the Likert scale, about the schemes in the AEP are shown to be related to the general attitude: schemes should be voluntary (negative relation), they should be administered by governmental organisations and contracts should be valid for a longer time period.

Concerning the expected effects on farm production, all signs for the coefficients are as expected and the relations are significant. Likert scores increase with the expectation that production or profit increases and that work or investment decreases.

Table 5.6
Regression model for EU8 with average Likert score as dependent variable

Predictor	β-coefficient	Standard deviation	T-value	P-value
Intercept	0.330	0.013	24.47	0.000
French-dummy	0.216	0.009	21.91	0.000
Austrian-dummy	0.092	0.009	9.31	0.000
German-dummy	0.090	0.011	7.78	0.000
Greek-dummy	0.067	0.011	5.83	0.000
Italian-dummy	0.121	0.010	12.02	0.000
Swedish-dummy	0.033	0.009	3.40	0.001
British-dummy	0.101	0.014	7.13	0.000
Ranks environmental issues highest	0.015	0.004	3.25	0.001
Participation in AEP	0.032	0.005	5.53	0.000
Participation in preservation of landscape appearance	−0.009	0.005	−1.67	0.096
Participation in preservation of biological diversity	0.013	0.006	2.26	0.024
Participation in reduction of agri-chemicals	0.011	0.005	1.97	0.049
Neighbour or relative applied AEP	0.013	0.005	2.37	0.018
AEP will increase farm production	0.006	0.003	1.74	0.082
AEP will increase profits	0.019	0.003	5.88	0.000
AEP will increase working efforts	−0.008	0.003	−2.47	0.014
AEP will increase investments	−0.011	0.003	−2.95	0.003
AEP should be voluntary	−0.013	0.005	−2.41	0.016
AEP should be administered by governmental organisations	0.013	0.004	3.31	0.001
Length of AEP-contracts (in years)	0.017	0.003	5.59	0.000
Has children less than 18 years old	−0.006	0.003	−1.66	0.098
General education level (in years)	0.002	0.001	2.79	0.005
Profits increased previous five years	0.016	0.002	5.94	0.000
	F-value	P-value	R²(adj)	
	87.07	0.000	0.545	

AEP is agri-environmental policy

In Table 5.7 the signs for the β-values for all the multiple regression models are presented. The columns represent models run on nine different data sets. The explanatory power of the models differs from a R^2 (adjusted)-value of 0.55 for the whole data set to 0.20 for the models of Sweden and Italy. Also the F-values, which are an indication of the quality of the model, vary between the whole data set and the individual countries.

The main results are that less variables are significant, due to smaller samples, but that most of the signs for the individual countries coincide with those for the whole data set. Only for two variables, and in each case only for one country, the sign differs from that of the whole data set. This shows that the results for the EU8 are quite robust in relation to, for instance, selection of countries and regions. The following variables receive the same sign for at least four countries as for the whole data set: participation in AEP, expectation that AEP increase production; expectation that AEP increase profits, AEP should be administered by governmental organisation, wants longer AEP contracts and profits increased during the last five years. As can be noted, there are several country differences. The limited number of variables that are significant can be explained by the small samples but the difference in signs for farm characteristics show that there are real country differences. It should be interesting for future research to look further into this. A more complete analysis of the country data is given in Bergström et al. (1999).

5.4.3. Model explaining participation

The household utility model of farm households, based on neo-classical economic theory, has the objective of structuring the factors that may affect farmers' behaviour in relation to willingness to participate in the AEPs. The dependent variable is participation in at least one AEP. The main factors that are expected to affect behaviour are: expected change in profit, expected change in labour and own satisfaction caused by improvement in landscape amenities. The hypothesis for these important variables are supported by empirical evidence. We also believe several factors such as farm type, farm size, age may affect the probability of participation.

In this group of regression models, all farmers who are not eligible to enter the farm into any scheme are excluded, since those farmers have not been able to choose whether or not they wish to participate in AEP. The responses are not weighted since the software that was used cannot handle a weighted logit-model. In the model described in Table 5.8, 1247 responses were used and 750 responses were rejected due to missing values for at least one variable.

Farmers in France and Sweden tend to have a higher and in Germany a lower participation rate than the average. This result is a confirmation of Hypothesis: Country, i.e. that there are differences among countries.

Farmers who expect production or profit to increase have a high participation rate. This is a confirmation of Hypothesis: Economic. A high ranking of environmental preservation, i.e. first, second or third and the factor of economic Likert statements tend to have a positive influence on participation. The Hypothesis: Attitude is thereby confirmed.

Number of employees, which is closely linked to the size of a farm, tends to increase participation, thus confirming Hypothesis: Size. Previous participation in other environmental programmes has a positive influence on uptake, which was expected according to Hypothesis: Experience. Years of general education enhances enrolment in the programme, which implies a confirmation of Hypothesis: Education.

Farmers are also affected by neighbours and relatives who have applied for a contract, thus confirming Hypothesis: Neighbour. The age of the farmer also influences participation but in the opposite way as compared to the hypothesis. Hypothesis: Age is thus rejected.

The opinions "that contracts should be extended in time" and "being well informed about the AEP" are related to participation. Those who own most of the land used on the farm and those who expects the farm to be abandoned also tend to participate to a larger extent than others. The first is what one can expect, but the latter result is somewhat surprising.

Note that while participation by relatives and neighbours is a significant explaining variable, being encouraged by friends and relatives was only ranked as the seventh most important reason to participate.

Table 5.7
Signs of independent variables for all models explaining attitude

Predictors	EU8	France	Austria	Germany	Belgium	Greece	Italy	Sweden	UK
Environment preservation highest ranked of public issues	+								−
Environmental policy high rank of public issues (1, 2, 3)	+	+		+		(+)			+
Participation in AEP	−								
Participation in preservation of landscape appearances	+	+		+	+		+		+
Participates in preservation of biological diversity	+					+			
Participates in programmes for ESA's				−					
Participation in programmes for the reduction of agri-chemicals	+								
Participates in programmes for organic production									(−)
Promotion of extensive and traditional production									
Promotion of set-aside			+						
Participates in other type of AEPs			−						
Well informed about AEPs									+
Neighbour or relative has applied for AEP-contract	+					(+)			
AEP will increase farm production	(+)	+		+		+			
AEP will increase profits	+		+	(+)	+			+	+
AEP will increase working efforts	−					−			
AEP will increase investment demand	−		−		(−)				(−)
Previous participation in other conservation programme	−								
AEP should be voluntary	−				+				
AEP should be administered by government organisation	+				+			+	+
How long should the AEP-contracts be (in years)	+		+					+	
Compensation for conservation should be based on effort					+				

	(1)	(2)	(3)	(4)	(5)	(6)	(7)	(8)	(9)
Respondent's age		+				+	(+)		–
Has children less than 18 years old	(–)								
General education level (in years)	+	(–)				–		–	(+)
Agricultural education level (in years)			–			–			–
Farm income (log(Euros/year))						–			
Farm income as proportion of household income						–			
The profits have increased during last five years	+	+			+		+	+	+
Farm income is supplemented by public funds			–			–			
Number of employees (including family)							(–)		
Number of animals (log)					(–)				
Number of hectares (log)			+	–		+			
Respondent's farm is an animal farm				(–)					
Owns most of the land in use								–	+
Experience of farming (in years)				(–)	(+)	–			+
Expects that the farm will be inherited		–					+	+	
Expects that the farm will be abandoned		–							–
N	1654	187	164	177	245	236	226	161	236
F-value	87.1	5.0	7.0	9.4	8.7	8.1	12.0	6.6	7.8
R^2(adj)	0.55	0.19	0.23	0.25	0.29	0.31	0.20	0.20	0.32

Signs without parenthesis represent significance at the 95% level and with parenthesis at the 90% level.
AEP is agri-environmental policy and ESA is environmentally sensitive area.

Table 5.8
Logit model of EU8 explaining participation without farm income as an independent variable

Predictor	β-coefficient	Standard deviation	Z-value	P-value	Odds Ratio
Intercept	−4.944	0.593	−8.34	0.000	
French dummy	0.988	0.227	4.35	0.000	2.69
German-dummy	−0.828	0.289	−2.87	0.004	0.44
Swedish-dummy	0.769	0.296	2.60	0.009	2.16
Ranks environmental issues high	0.386	0.154	2.51	0.012	1.47
Well informed about AEP	1.677	0.156	10.78	0.000	5.35
Neighbours or relative has applied	0.857	0.155	5.53	0.000	2.36
AEP will increase farm production	0.398	0.148	2.68	0.007	1.49
AEP will increase farm profits	0.429	0.134	3.21	0.001	1.54
Participated in environmental programme	1.312	0.251	5.23	0.000	3.71
Length of contracts (in years)	0.234	0.113	2.07	0.039	1.26
Factor of economic Likert statements	2.250	0.403	5.58	0.000	9.49
Respondent's age	0.018	0.007	2.48	0.013	1.02
General education (in years)	0.048	0.028	1.71	0.087	1.05
Number of employees (incl. farmer)	0.067	0.042	1.61	0.108	1.07
Owns most of the land in use	0.452	0.162	2.80	0.005	1.57
Expects farm will be abandoned	0.970	0.486	1.99	0.046	2.64
Participate	651				
Does not participate	596				
Log-Likelihood	−575.47				
P-value	0.000				
Concordant classification	86.3%				

AEP is agri-environmental policy.

If farm income is included as an independent variable in the model, the number of responses that can be used are reduced, i.e. we gain information in one respect and loose in another. In this model, shown in Table 5.9, 820 responses were included and 1177 responses were rejected due to missing values for at least one variable.

Farm income did not turn out to be significant, but has the expected sign. The model is about as good as the model without income as a dependent variable. Most variables that are significant in the first model are also significant in the second model. Two variables are significant only in the first model, i.e. AEP will increase production and expects the farm will be abandoned. Only one variable, i.e. high scores on general Likert statements, turns out to be significant in the second model but not the first. The second model is largely a confirmation of the first.

For EU8 the Hypothesis: Income, Hypothesis: Animal and Hypothesis: Age were rejected. The other seven hypotheses have been confirmed in the tests, and some are also supported by other facts.

Table 5.9
Logit model of EU8 explaining participation with farm income as a predicting variable

Predictor	β-coefficient	Standard deviation	Z-value	P-value	Odds Ratio
Intercept	−11.434	1.510	−7.57	0.000	
French dummy	1.2948	0.3421	3.79	0.000	3.65
German-dummy	−0.7765	0.4150	−1.87	0.061	0.46
Swedish-dummy	2.3438	0.4566	5.13	0.000	10.42
Ranks environmental issues highest	−0.6338	0.3278	−1.93	0.053	0.53
Ranks environmental issues high	0.6781	0.2182	3.11	0.002	1.97
Well informed about AEP	1.9106	0.2073	9.22	0.000	6.76
Neighbour or relative has applied	1.1191	0.2087	5.36	0.000	3.06
AEP will increase farm profits	0.7671	0.1411	5.44	0.000	2.15
Participation in other environmental programmes	0.8712	0.3346	2.60	0.009	2.39
Length of AEP-contribution (in years)	0.4622	0.1512	3.06	0.002	1.59
Factor of economic Likert statements	1.7841	0.5392	3.31	0.001	5.95
Factor of general Likert statements	2.1863	0.7787	2.81	0.005	8.90
Respondent's age	0.0282	0.0098	2.88	0.004	1.03
General education (in years)	0.1221	0.0372	3.28	0.001	1.13
Farm income (log(Euro/year))	0.3005	0.2487	1.21	0.227	1.35
Profits increased previous five years	−0.3375	0.1416	−2.38	0.017	0.71
Income also from public funds	1.6882	0.4623	3.65	0.000	5.41
Animal farm	0.3438	0.2453	1.40	0.161	1.41
Log of number of hectares	0.4557	0.1854	2.46	0.014	1.58
Owns most of the land in use	0.3981	0.2190	1.82	0.069	1.49
Participate	446				
Does not participate	374				
Log-Likelihood	−350.400				
P-value	0.000				
Concordant classification	88.5%				

AEP is agri-environmental policy.

In Table 5.10 the signs for all β-values of the logit regression models are presented. The columns represent models run on nine different data sets. The main result is that less variables are significant, as compared with EU8 due to smaller samples, but that most of the signs for the individual countries coincide with those for the whole data set. The following variables receive the same sign for at least three countries as for the whole data set; participating neighbours or relatives and high scores on economic Likert statements. There are important country differences especially concerning variables on farm characteristics. A more complete analysis of the country data is given in Bergström et al. (1999).

Table 5.10
Signs for independent variables for all logit models explaining participation

Predictors	EU8	EU8 including income	France	Germany	Belgium	Greece	Italy	Sweden	UK
French-dummy	+	+							
German-dummy	−	(−)							
Belgian-dummy									
Greek-dummy									
Italian-dummy									
Swedish-dummy	+	+							
British-dummy									
Environmental policy highest rank of public issues		(−)							
Environmental policy high rank of public issues (1, 2, 3)	+	+			+	+			
Well informed about AEP	+	+	+		+				+
Neighbour or relative has applied for AEP-contract	+	+	+		(+)	+			+
AEP will increase farm production	+				+				
AEP will increase profits	+	+	+	(−)		+			
AEP will increase working efforts					−				
AEP will increase investment demand									
Has previously participated in other environmental conservation programmes	+	+		+	+				
AEP should be voluntary			(−)						
AEP should be administered by governmental organisation	+	+							(−)
How long should the AEP-contracts be						+			(+)
Compensation for conservation based on effort			−	−					
Factor of environmental Likert statements			+	+	+				
Factor of economic Likert statements	+	+					+	+	
Factor of administrative Likert statements						+	+		
Factor of general Likert statements	+	+			(+)				+

Respondent's age	+			(−)					+
Has children less than 18 years old	(+)	+							
General education level (in years)		+		(−)					
Agricultural education level (in years)					+				+
Household income (log(Euros/year))									
Farm income (log(Euros/year))					+	+			
The profits have increased last five years		−				+	+		−
Farm income is supplemented by public funds		+						+	−
Number of employees (including family)			+	+					
Number of animals (log)		+						+	+
Number of hectares (log)			−			+			
Respondent's farm is an animal farm			(+)						−
Owns most of the land in use	+	(+)				+			
Experience of farming (in years)							(+)		
Expects that the farm will be inherited								+	−
Expects that the farm will be abandoned	+							+	
N	1247	820	284	164	246	243	209	183	223
Concordant-classification	86%	89%	87%	91%	88%	88%	93%	88%	89%

AEP is agri-environmental policy.

5.5. Summary and conclusions

The following discussion of attitude to and uptake by farmers of AEP is based on a rather large survey ($N = 1997$) within nineteen regions in the eight STEWPOL countries. Although, the sampling and interviewing procedure have not been exactly identical in the eight countries due to lack of access to farm registers in most of the countries, we believe that the results represent attitudes and uptake behaviour in each of the sampled regions sufficiently well. The survey is not representative for the EU as a whole, since only parts of the EU are represented, nevertheless the survey reveals an interesting and relevant indication of attitudes and uptake behaviour in the EU. In general, the econometric models are good predictors and many variables are found significant at the compiled EU8-level.

In general the hypotheses, that were largely based on a theoretical discussion, were confirmed by statistical tests and/or supported by other evidence. The most obvious result from the survey as a whole, is that expected change in profits matter. The simple policy conclusion is that an increase in support levels would increase participation. This is technically easy to do, but EU and national budgets have alternative uses so there must be large enough expected social benefits from environmental improvement to mitigate the loss elsewhere in the economy from using more of scarce budget funds for this purpose. The relatively large proportion of participants who state they would do the same or almost the same, whether they receive the compensation or not, should only be seen as relevant in the short run.

Use of own labour time, with opportunity cost in on and off farm work or recreation, is another factor farmers seem to consider when deciding to participate or not. In this case, policy conclusions may be less obvious since most of the programme include compensation for calculated (or estimated) labour cost. It should be noted that farmers own transaction costs may be considerable. For complex policies, concerning for instance maintaining or enhancing biodiversity, these costs are estimated to be around 15% of the compensation paid in Sweden (Kumm and Drake, 1998). If full compensation for production or transaction costs is not paid, many farmers are likely to choose not to participate.

Also, farmers' attitudes towards the environment matter. This can be interpreted as if farmers' attitudes affect their own satisfaction from an improved environment, but an interpretation of this as altruistic behaviour is also possible. The authorities can try to influence behaviour by information campaigns aiming at attitudinal changes. This is also related to the importance of information which is discussed below. But the most direct solution may be to change some policies so that the environmental impacts are more clear and are accepted by farmers.

The finding that participating neighbours or relatives have a positive influence on own participation, speaks in favour of targeting the marketing of policies towards farmers who are normally more interested in new ideas. Many of their neighbours are likely to follow in due course. The fact that neighbours tend to have similar conditions may, however, explain some of the correlation.

Years of general education have a positive influence on attitude and participation and lack of information is a main motive for not participating. This indicates that efforts to improve knowledge and information would increase participation. This may turn out to be cheaper than an increase in compensation levels.

Those who are positive to long term contracts have a higher participation rate than others, probably because most contracts run for 5 years. In order to increase participation among farmers who wish to be less constrained, the AEP may need to be more flexible in relation to contract time. This is, of course, only relevant when environmental performance does not necessitate long term contracts.

The size of farms, measured both in hectares and in number of farm workers, has a positive influence on participation. This may be caused by several factors. In some cases, the rules of the AEP may make it difficult for very small farms to become eligible and there may be economies of scale in producing the environmental benefits and in transaction costs. Also the marginal effect of lower production may be less important for larger farms than for smaller. The solution can be to pay a small part of the total compensation to each participating farm instead of in relation to quantity of work or hectares involved.

Overall the study has shown that farmers in the eight EU countries are in general positive towards the present agri-environmental support schemes and that there are a number of ways to increase participation. In short:

1. increased payouts, including remuneration for transaction costs;
2. improved information about the schemes, information should especially appeal to farmers environmental concern and conscientiousness;
3. improved targeting of schemes on environmental improvement;
4. targeting information on the more advanced farmers;
5. increasing the flexibility in relation to contract time;
6. special support to encourage small farms to participate.

References

Becker, G. (1973) A Theory of Marriage: Part I, *Journal of Political Economy*, 81(4), 813–846.

Bergström, P., Svedsäter, H. and Drake, L. (1999) *Farmers Attitudes to and Uptake of Countryside Stewardship Policies*. Final report of the EU-project CT95/0709, Swedish University of Agricultural Sciences.

Eagly, A. H. and Chaihen, S. (1993) *The Psychology of Attitudes*. Harcourt. Fort Worth, TX, USA

Gujarati, D. N. (1988) *Basic Econometrics*. 2nd Edn, McGraw-Hill International Editions, Economics Series, Singapore.

Kumm, K.-I. and Drake, L. (1998) *Transaction Costs to Farmers of Environmental Compensation*. Department of Economics, Swedish University of Agricultural Sciences, Uppsala, Sweden, (unpublished paper).

Lexmon, Å. and Andersson, H. (1998) Adoption of Minimum Tillage Practices. Some empirical evidence, *Swedish Journal of Agricultural Research*, 28, 29–38.

Newcomb, T. M. (1953) *A Theory of Cognitive Dissonance*. Rowman-Peterson, Evanston, IL.

Robson, C. (1993) *Real World Research*. Blackwell Publishers, Oxford, England.

Singh, I., Squire, L. and Strauss, J. (1986) The Basic model: The Theory, Empirical Results and Policy Conclusions. In: I. Singh, L. Squire and J. Strauss (Eds), *Agricultural Household Models—Extensions, Applications and Policy*. The John Hopkins University Press, The World Bank, Baltimore, USA.

G. Van Huylenbroeck and M. Whitby
Countryside Stewardship: Farmers, Policies and Markets
© 1999 Elsevier Science Ltd. All rights reserved

Chapter 6

Modelling regional production and income effects

Ottmar Röhm and Stephan Dabbert

Abstract—This chapter focuses on the analysis of regional production and income effects of CSPs, in order to give some answer to the question whether CSPs can contribute to market equilibrium. Closely linked with the income effects of CSPs is the complex of the profitability of cultivating marginal land and the land abandonment issue, which itself has obviously strong impact on the production topic.

The analysis is based on model calculations in eight EU regions (NUTS II) using the technique of Positive Mathematical Programming. The chapter starts with an overview on the applied methodology, followed by a section providing a more extended analysis of the results obtained for one study region. The final section summarises the general results and draws some conclusion.

6.1. Introduction

Much of the interest in agricultural policy is still centred around production and income effects in spite of the growing importance of environmental objectives. Because of this it is important to consider the impact of CSPs on these more traditional variables of interest of agricultural policy. In order to fully evaluate CSPs, their effects on production quantity is of major importance. If CSPs would lead to a decrease in production quantity this would certainly be an argument in favour of them, at least as long as the situation with surplus production in the European Union prevails.

In order to investigate whether or not CSPs do have that effect it is necessary to use a modelling approach. The necessity to use such an approach becomes clear if one tries to imagine how to empirically investigate this question. In reality, at the same time as the introduction of CSPs, many other important variables also changed, e.g. prices and technologies. Given this fact there is no easy way to attribute any change in production quantity to the introduction of environmental policies in agriculture. This scientific problem of empirical investigation is aggravated by the fact that most CSPs have only been implemented for a very short time and in most cases acceptance rates are low so that only small production effects could be detected in any case.

The modelling approach helps to find a way out of this dilemma. This chapter is based on mathematical model building. The advantages of such modelling include the following: mathematical models are able to represent clear cause–effect relationships and scenarios of situations that have not yet been observed. It is thus much easier to isolate the effect of specific CSPs within the model than it is in reality. For the economist a mathematical model can be regarded as a tool that is both artificial and useful—like the laboratory of the natural scientist. However, as there are major differences between the laboratory and reality, so there are between the mathematical model of a part of the economy and reality. For this reason it is important to stress the need for a very cautious and careful interpretation of any quantitative results coming out of mathematical models. Actually in many cases the experience gained by the modeller in building and operating the model is as valuable as the quantitative modelling results themselves. The reason for this is that the exercise of building a mathematical model forces the modeller to think very carefully through and analyse the different parts of the system investigated and their interconnections. The modeller is also forced to take a clear look at the limitations of his model, as all models—by their very nature—leave out some aspects of reality. For these reasons a qualitative verbal interpretation of the results is the most important output of the modelling exercise.

In the following part of this chapter we will first describe important aspects of the methodology used for mathematical modelling. We will then present some results and draw conclusions.

6.2. The methodological approach

6.2.1. Choosing the policies to be investigated

The large number of CSPs (also Chapters 2 and 3) implemented in the participating countries made it necessary to limit the scope of policies for the quantitative analysis. The analysis is focused on policies that were implemented under Regulation 2078/92 (Deblitz and Plankl, 1998). In addition the policies have to be either applied on a relevant acreage or have to be combined with relevant transfers to farmers. It is obvious that measurable effects on an aggregated level can not be expected from measures covering only a few hectares.

The markets under consideration are those of the main arable crops and the livestock sector, whereas permanent crops are not analysed in detail. Environmental effects of CSPs and the production of stewardship goods and services are not modelled explicitly in the cross-country approach reported here. There are other attempts to include these effects in regional models on a more limited regional scope (e.g. Dabbert *et al.*, 1999). However, the qualitative result of such studies is that an extensification of production is causing a reduction of negative externalities of agricultural production (Frede and Dabbert, 1999), whereas the upkeep of marginal farm land is usually combined with the maintenance of positive externalities.

The CSPs can only be analysed adequately if the general policy framework in which they act is also represented and its relevant parts modelled explicitly. This is mainly due to the fact that the monetary transfers associated with these general policies (LFA payments, also some general CAP policies, like arable support scheme, livestock support and beef extensification scheme) frequently outweigh those of other sources (like CSPs) and therefore determine production decisions to a large degree.

In the modelling 'laboratory' the following approach was used to isolate the effects of CSPs: The model was used to depict the situation in the reference year (1995) on the basis of the policies implemented at that moment, including CSPs. Due to the modelling approach chosen, which will be explained hereafter, the model exactly describes the situation which has actually been observed.

Then, in a second step of the modelling experiment, all compensatory payments for the CSP activities were set to zero while the farmers could still apply the specific CSP production techniques if it is financially viable for them. If this is the case, it means that in a world of 1995 without state payments for CSPs would have existed, some farmers would still have applied CSP technologies. Only the additional amount of acreage induced to take up CSP by the introduction of premiums actually may be called a policy effect. With this approach the model is therefore able to measure exclusively those effects which are motivated by the CSP schemes.

6.2.2. Choosing the appropriate type of model

There are several instruments available for agricultural policy analysis (Bauer and Henrichsmeyer, 1989), so the first step is to select the most suitable model. The two most important aspects here are the mathematical structure (econometric versus programming model, linear versus non-linear) and the aggregation level (farm level, regional level, sector level) of the analytical model.

Given the limited experience with CSPs and the absence of long-term time-series data, econometric models are not an appropriate tool, thus programming models are the best approach to the problem. In general the strength of a programming modelling approach lies in its ability to include clear-cause relationships and to analyse situations that could not yet be observed.

Most of the CSPs are implemented at a regional level. This means that this is the highest possible aggregation level. Another possible level of analysis would however be farm level case studies (see Deblitz, 1998). The main strengths of the regional approach lie in the possibility for a broader spatial coverage, and in its ability to assess production effects directly at a more aggregated level. Of special importance is also that the use of the regional level allows it to base the analysis of CSPs on empirically observed participation rates. However, the regional approach is subject to the so-called aggregation error which is its main disadvantage. The aggregation error is caused by hyper-optimal resource allocation, in comparison to reality (see Day, 1963).

The conventional programming model is a linear model where both objective function and restrictions are linear. Though powerful and thus widely applied in agricultural economics through the last decades this approach has several shortcomings. One of the major shortcomings of traditional LP models applied on a regional level is their tendency to lead to overspecialised results compared to the empirically observed situation. The main reason for this overspecialisation is the lack of empirically justifiable constraints that can be used to construct the model. The importance of these constraints is given by the mathematical fact, that the number of activities in the solution vector equals at the most the number of binding constraints. The problem to specify a sufficient number of constraints appears especially at regional and higher aggregated levels. Here the problem is that some potentially available resource constraints are difficult to measure (due to time or budgetary constraints), as is the case for example with stable capacities, or with the classification of soil categories with different quality levels. Other resource constraints, for example labour become useless due to aggregation error. These problems are the more severe the more heterogeneous the regions. In order to reduce the problem of overspecialisation and to calibrate the models to the reference situation, so-called calibration constraints are implemented in LP-models. However these calibration constraints might predetermine the scenario results in an unreasonable way.

A further disadvantage of LP-models is that they respond discontinuously to changes in external variables, e.g. a constantly rising premium for a certain CSP might first lead to no reaction of the model over a certain range and then suddenly to a dramatic change in the production structure.

6.2.3. Positive mathematical programming

The innovative methodology of positive mathematical programming (PMP) has been developed by Howitt (1995) to overcome the major weaknesses of Linear Programming. It is especially suited for regional applications and has the capacity to overcome the calibration problem to a large degree, without limiting the model flexibility by calibration constraints. The general idea is to replace the linear objective function of a programming model by a non-linear function. This non-linear function, in most cases a quadratic function, is derived from empirically observed variables from a baseline period in a specific way. Therefore the method is also described as Positive Quadratic Programming (PQP).

The basic assumption for the methodology is that farmers behave optimally in an economic sense, and adapt their production structure to the general agri-political framework. The economic condition for optimal allocation is that marginal gross margins are equal for all production activities with respect to a scarce resource.

In this paragraph the basic steps to build a PQP model are described very briefly, for this the following abbreviations are used:

X_i: activity level (X) of different crops (i)
$\hat{X}_i$: activity level at the reference situation
y_i: yields p_i: product prices P_i: CAP premiums
vc_i: variable cost GM_i: gross margins

In a first step a linear programming model is constructed in which the cropping structure of the reference year is represented exactly with the help of calibration constraints. This model is not suited for scenario calculations, because it is overly restricted by the calibration constraints and can therefore barely react to changes in external variables. However, it provides the dual values for the resource constraint (arable land λ_L) and for the calibration constraints (λ_i), with which the coefficients of the quadratic objective function can be calculated.

The LP model has the following structure:

$$\max f(X)$$

with $\quad f(X) = TGM = \sum_i (GM_i * X_i) = \sum_i [(y_i * p_i + P_i - vc_i) * X_i]$

subject to:

$$\sum_i X_i \leq \sum_i \hat{X}_i = R \qquad \text{resource (land) constraint} \quad \lambda_L$$

$$X_i \leq \hat{X}_i * (1 + \varepsilon)^1 \qquad \text{calibration constraint} \qquad \lambda_i$$

$$X_i \geq 0$$

The dual values for the calibration constraints (λ_i) show how much gross margin is being lost (at the margin, that is for the last unit) because of the calibration constraint. In a truly profit maximising equilibrium all marginal gross margins of all realised land use activities should be equal. Thus, in such a situation there would be no incentives to change the level of any activity meaning that marginal revenue equals marginal cost. It is obvious that then the marginal gross margin in the LP model is overestimated by the magnitude of the dual value of the binding calibration con-

---

[1] ε is a very marginal figure, necessary to avoid uncertain distributions of dual values.

straint. There are two possible reasons why the marginal gross margin in the LP solution is higher than it should be in the profit maximising equilibrium: either marginal variable cost is underestimated or marginal yield is overestimated.

In fact, the PQP model can principally be calibrated with two different approaches, the first one based on the assumption of increasing marginal cost, the second one based on the assumption of declining marginal yields. The explanation here refers to the second approach, in the actual model both approaches were combined.

Assuming a simple quadratic form, the general structure of the objective function of a PQP model, calibrated on declining marginal yield is:

$$TGM = \sum_i [(y_i * p_i * (\alpha_i - \gamma_i * X_i) + (P_i - vc_i)) * X_i]$$

With the above information on underestimated marginal cost, or overestimated marginal yields we can recalculate the marginal gross margin at the empirically observed point ($\hat{X}_i$): It is the average GM_i minus the dual value of the calibration constraint. This allows us to set up a first equation to calculate the coefficients α_i and γ_i:

$$MGM_i = \frac{\partial TGM}{\partial X_i} = y_i * p_i * (\alpha_i - 2 * \gamma_i * \hat{X}_i) + (P_i - vc_i) = GM_i - \lambda_i = \lambda_L \qquad (1)$$

The marginal gross margin of each crop at the empirically observed point corresponds with the dual value of the land constraint.

To determine the two coefficients α_i and γ_i a second equation has to be set up. It can be easily derived if one demands that the average gross margin (AGM) for each crop in the PQP model up to the observed activity level has to be equal with the linear GM of the crop:

$$AGM_i = y_i * p_i * (\alpha_i - \gamma_i^* \hat{X}_i) + (P_i - vc_i) = GM_i \qquad (2)$$

After some simple transformations one obtains:

$$\alpha_i = 1 + \frac{\lambda_i}{y_i * p_i}$$
$$\gamma_i = \frac{\lambda_i}{y_i * p_i * \hat{X}_i}$$

A general difficulty in the calibration process is that by definition the dual value of the calibration constraint for the least profitable crop is zero in any case. With a dual value of zero, the objective function for this crop stays linear in principal (that can easily be derived from the formulas above), which leads to problematic implications with respect to model behaviour. However, it is possible with some information on the actual shadow price of land to create a non-linear objective function also for the least profitable crop, whereas the principal way of calculating the coefficients is not changed. For further information on this issue see Howitt (1995). After calculating the coefficients one can construct the PQP model, which is of following structure:

max $f(X)$

with $f(X) = TGM = \sum_i [(y_i * p_i * (\alpha_i - y_i * X_i) + (P_i - vc_i)) * X_i]$

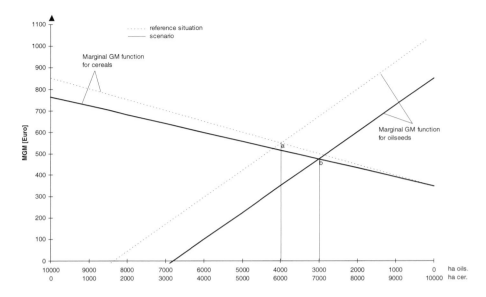

Fig. 6.1. **Reaction of the model to changes in external variables**.

subject to:

$$\sum_i X_i \leq \sum_i \hat{X}_i = R \qquad \text{resource (land) constraint}$$
$$X_i \leq 0$$

The model works now without any calibration constraints. It can be shown that the model calibrates exactly to the reference situation. This model is now suited for scenario calculation.

Figure 6.1 illustrates graphically the reaction of a PMP model to changes in external variables. The figure is based on an hypothetical example.

The changes in external variables are likely to represent an Agenda 2000 scenario. For cereals a price reduction and an increase of the direct arable payments is assumed and for oilseeds just a reduction of the direct arable payments. The figure shows the graphs of the marginal gross margin functions in the reference situation and under scenario conditions. The optimal allocation is marked by the respective points of intersection of the graphs.

Point **a** marks the cropping structure in the reference situation, whereas point **b** shows the new optimal resource allocation under scenario conditions with changed external variables. One can see that the acreage of cereals is extended whereas the area cultivated with oilseeds is reduced correspondingly. The dual value of land declines, and also the total gross margin which is represented by the area under the respective graphs.

The way the graph reacts on changed external parameters is also interesting. Price changes lead in a PQP model based on declining marginal yield to a revolving of the graph of the MGM-function, whereas in case of premium changes the graphs are shifted in a parallel fashion.

The description given here provides only a rough overview on the general methodology. For the actual applied models the methodology needed further development, especially to include the CSP activities and to link the livestock sector with the forage production, as represented by the graphical overview of the actual model structure in fig. 6.2.

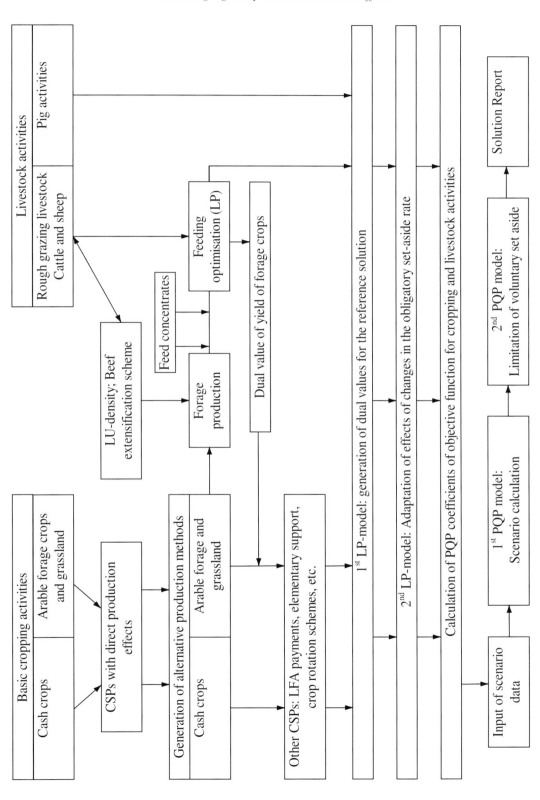

Fig. 6.2. **Principal, simplified structure of the analysis model.**

Table 6.1.
Main characteristics of CSPs in the study regions

Region	Country	Code	Regulation 2078/92 transfers		Grassland		Arable land	
					Major focus of schemes			
			MECU	ECU/ha UAA	'upkeep'	'extensify'	'upkeep'	'extensify'
Oberösterreich	Austria	AT31	94.0	164	+	+	+	+
Luxembourg	Belgium	BE34	0.1	1		+		
Oberpfalz	Germany	DE23	15.2	37	+	+	+	+
Thessalia	Greece	GR14	5.8	12				+
Auvergne	France	FR72	40.1	26	++			
Veneto	Italy	IT32	8.9	10		(+)		+
Sydsverige	Sweden	SE04	13.3	22	+	(+)		(+)
Cumbria	UK	UK12	2.7	6		+		

6.3. Case studies of 8 European regions

The analytical models, based on the PMP methodology, have been applied to eight different study regions to analyse the effects of CSPs on regional farm income and production. These regions are located in eight STEWPOL countries and represent a vast array of different situations with respect to agricultural policy (especially CSPs) and natural conditions for agricultural production. These regions, though not representative for all of Europe in a statistical sense, give a good impression of the situations that might also be expected in other regions in Europe. As the further development of CSPs in Europe will be decisively influenced by Agenda 2000 policies an analysis of potential effects of Agenda 2000 has also been done, which will be referred to in a general way without presenting quantitative details. Table 6.1 shows the main characteristics of CSPs in the sample regions. Among the numerous schemes two main groups or types of schemes have been identified. The classification is very crude but it shows the main focus of the schemes with respect to the topic of the analysis done here.

The first group of measures addresses mainly the *upkeep of agricultural land use* (termed 'upkeep' in Table 6.1). In this group is the French '*Prime à l'herbe scheme*', the 'elementary support schemes' of Austria and Oberpfalz and most of the Swedish schemes, which aim at the preservation of semi-natural grazing land. These schemes do not have major extensification effects (mainly because farming is already very extensive in the regions where these measures are implemented). Therefore no major market effects, in the sense that output is reduced, can be expected. They mainly stabilise farm income and (shadow) prices of land. However these policies are nevertheless quite different from general CAP support instruments. In contrast to other CAP policies these CSP schemes are more focused, by often excluding intensive farms on favourable sites.

The second group of schemes consists of measures which lead actually to an *extensification of production* (termed 'extensify' in Table 6.1), combined with decreasing yields or stocking densities per hectare. These measures often require special management techniques, or ban certain yield

Table 6.2
Farm statistical data on the sample regions (1995)

Region	Quality[1]	Territory	UAA	Arable land	Grassland[2]	Permanent crops	Cattle	Sheep	Pigs
				[1000 ha]			[1000 heads]		
AT31	+	1198	574	295	257	0	694	51	1180
BE34	−	444	139	21	118	0	384	10	14
DE23	0	969	417	291	126	0	543	33	296
GR14	+	1404	501	416	n.a.	60	76[3]	1405[3]	183[3]
FR72	−	2617	1521	307	1209	5	1523	789	279
IT32	+	1836	878	641	130	106	1058	32	546
SE04	+	1397	595	506	89	0	296	45	821
UK12	−	681	448	24	423	0	540	2663	66

[1] Natural conditions in comparison to the rest of the country: (+ indicates an above country average, 0 indicates approximately average, − indicates below average).
[2] Including temporary grassland.
[3] 1993.

increasing inputs. One of the most important schemes in this group is organic farming. If farmers are participating in such schemes at a large scale, market effects can be expected. Such schemes are often horizontal with uniform premium payments, so they include also an income support element. The less favourable the site on which farmers apply these measures, the higher is the share of income support.

The selection of the study regions has been performed according to several objectives. The selected regions were intended to represent major aspects of European farming, that means the relevant farm types as well as the broad variety of natural conditions has been considered. Table 6.2 provides some farm statistical data on the land use structure and the main livestock categories of the sample regions. It also gives a short overview on the quality of the respective region with respect to agricultural production compared to the rest of the respective country.

In relative terms (compared with the respective countries) the result is quite balanced between more and less favourable farming conditions. In absolute terms (compared to an absolute scale of natural conditions in the EU) the picture is somewhat different because very favourable sites are not included. The reason for this is simply that in the most favourable regions there are very low participation rates in CSPs, so that a quantitative analysis of existing CSPs for these regions is less interesting.

Although the most important aspect for the selection of the sample regions was the highest possible participation in CSPs (within their respective country), there are countries and regions with a practically negligible acceptance and application of CSPs (see Table 6.1). In the Austrian study region 'Oberösterreich' CSPs are of outstanding importance, especially if one regards the premium transfer on a per hectare base, and in addition all types of CSPs are applied in this region. Therefore this region has been chosen to present the result of the modelling efforts in detail. The major results for the other regions are summarised in the subsequent section.

6.4. Oberösterreich—detailed modelling results for a European region that is a leader in CSPs

In general terms Austria is one of the leading European Countries concerning the implementation of CSPs under regulation 2078/92. Within Austria Oberösterreich has been selected, because it represents quite a range of natural conditions and has a considerable acceptance of CSPs. Because of the variety of CSPs applied in this region it is a particularly useful example where many interesting features of CSPs can be demonstrated quite clearly. For this reason the results for this region are presented in some detail, starting with a description of the region and the policies implemented, a short description of special features of the regional model and an extensive presentation of some results of the policy analysis.

6.4.1. Geography and climate, agricultural production and CSPs

The Austrian model region, Oberösterreich (Upper Austria, AT31), is located in the central north of the country. The total area covers about 1,400,000 hectares with 1,400,000 inhabitants. Fifty-six per cent of the territory is used for agricultural purposes, 37% is covered by forest. Oberösterreich has, in common with the whole of Austria, a broad range of heterogeneous soil, climate and topographical conditions, from favourable arable sites in the valley of the Danube, upon low mountain areas, to the Lime Alps. This variety in natural conditions finds its expression in the structure of agricultural production, with a broad variety of crops cultivated and also the whole range of animal production covered. Further information on this is provided in Table 6.2 and Table 6.7. Due to the large share of grassland livestock rearing is also important in Oberösterreich. About one third of Austria's cattle (beef and milk) and pigs is produced in Oberösterreich. The cattle herd includes 220,000 dairy cows and 53,000 suckler cows. While the suckler cow herd has rapidly grown during the last years, the number of dairy cows is declining proportionally. This is a result of the response of farmers to the policy change in 1995 due to the adoption of the CAP. Beef production is still dominated by intensive fattening of bull calves based on silage maize, but more extensive production is gaining importance. For about one quarter of all male fattening cattle a second special beef premium is paid which can serve as an indicator for a less intensive way of production.

The farms in Oberösterreich are quite small, the average farms size is only about 11 ha. Nearly two thirds of the 50,000 farms in Oberösterreich are run by part-time farmers.

The Austrian agri-environmental programme according to Regulation 2078/92 is called ÖPUL and has been implemented in 1995 the year in which Austria joined the EU. Many of the ÖPUL measures are modelled according to programmes that had been implemented before. In comparison to Regulation 2078/92 programmes in other countries, ÖPUL has the highest participation rate and more than 90% of the UAA in Austria are managed according to at least one CSP scheme. Many of the ÖPUL schemes can be combined with one another (e.g. the important scheme 'elementary support' with the 'crop rotation scheme'). A major reason for the high acceptance seems to be that agri-environmental schemes had been implemented on a large scale before Austria became an EU Member State. A brief description of the ÖPUL measures that are integrated into the model is given in Table 6.3 and the following sections.

In addition to schemes from ÖPUL, two policies from other policy areas are considered to be CSPs and therefore part of the modelling effort: LFA payments and beef extensification premiums are quite important in financial terms and also have effects on agri-environmental objectives. A major effect of LFA-payments is to support agricultural production in marginal regions which are in many cases regions where positive externalities of agricultural production are most obvious

Table 6.3
ÖPUL schemes — programmes according to Regulation (EEC) 2078/92

Name of scheme	Premium [ATS/ha]*		Area under policy [ha]		Type of scheme
	Arable land	Grassland	Arable land	Grassland	
Elementary support	650	700	257,600	246,000	Upkeep
Crop rotation premium	1,510		196,000		
Organic farming	4,500	3,000	9,400	20,500	Extensification
Ban of agri-chemicals (whole farm)	3,000	2,000	19,400	42,300	,,
Extensive bread grains	2,400		10,300		,,
Ban of certain inputs	800–2,500		101,060		,,
Erosion control	500–1,000		260		,,
Extensification of grassland		1,600–1,800		>100,000	,,
Alpine pasture support		880		4,800	Upkeep

* 100 ATS = 7.27 EURO

(a beneficiary of these externalities is considered to be the tourism sector). The beef extensification premiums can also contribute to the viability of farming in such regions. In addition the beef extensification premium (though not included in the agri-environmental measures of the CAP) can potentially induce an extensification of beef production in areas where this production has become highly intensive. In the following paragraphs the most important ÖPUL schemes are described.

In order to be eligible for the *elementary support scheme* farmers have to comply with the code of good fertilising practice and the livestock density on farm must not exceed 2.5 LU/ha (2.0 LU/ha from 1998 on).

The central restriction for the *crop rotation scheme* is that the share of cereals and maize on arable land must not exceed 75%, and in addition a minimum share of winter cover crops (catch crops or forage crops) has to be planted. The premium is differentiated according to the minimum share of these winter cover crops on arable land from 900 ATS/ha (>15% arable land with cover crops) up to 1.900 ATS/ha (>35%).

The aim of the *extensive bread grain scheme* is to produce cereals (wheat, barley, oats, rye) which are used for the food chain with reduced input quantities. The use of fungicides or growth regulators is forbidden and in addition to some other minor requirements crop specific limits of N-fertilisers apply. A maximum of 40% of arable land per farm is eligible for support from this scheme.

The scheme promoting the *ban on certain yield increasing inputs on single plots* (arable land) consists of a very detailed set of extensification options:

- ban on growth regulators (only cereals eligible, 800 ATS/ha; 85,700 ha)
- ban on growth regulators and easily soluble mineral fertilisers (only cereals eligible, 2000 ATS/ha; 5700 ha)
- ban on synthetic plant protection substances and easily soluble mineral fertilisers (2500 ATS/ha; 1300 ha)
- ban on fungicides (only cereals eligible, 800 ATS/ha; 7600 ha)
- ban on synthetic plant protection substances (1,400 ATS/ha; 760 ha)

Table 6.4
Income effects of CSP scenarios in Oberösterreich

	Reference (with ÖPUL)	No ÖPUL		No ÖPUL No premium for voluntary set-aside	
	mn ATS*	mn ATS*	% of base	mn ATS*	% of base
Total regional gross margin	10,264	9165	89	9154	89

* 100 ATS = 7.27 EURO

The *erosion control measure* consists of two different sub-measures: For mulch-seeding a premium of 500 ATS/ha is paid, and for a switch from silage maize to other arable forage crops or grassland a premium of 1000 ATS/ha is paid.

The *grassland extensification scheme* requires the banning of easily soluble mineral fertilisers and large-scale application of synthetic plant protection substances (PPS) on grassland. Premiums from 1600 to 1800 ATS/ha are offered if farmers enter more than 30% or 60% respectively of their grassland into the scheme. Marginal grassland, like alpine pastures, is not eligible for the scheme.

6.4.2. The structure of the regional agricultural sector model of Oberösterreich

The large number of different crops and the multitude of different production methods for each crop resulting from the numerous CSP measures offered finds its correspondence in the model structure. Twenty-two crops are combined with 12 ÖPUL-measures with direct production effects in 137 cropping activities. The model is supplemented by 12 grassland activities representing three types of grassland.

Animal production is represented by 10 cattle and sheep activities, and 2 pig activities. Apart from the 12 ÖPUL measures effecting production directly, two additional measures, 'elementary support' and 'crop rotation scheme' were integrated into the model. All important schemes of the CAP, like set-aside scheme, beef premiums, arable support scheme are modelled explicitly.

6.4.3. Effects of CSP implemented in Oberösterreich

In the following, two scenarios concerning the analysis of the Austrian ÖPUL programme are presented in more detail. During the modelling process it became clear that abolition of ÖPUL will have a significant (positive) effect on voluntary set-aside of arable land, and therefore quite logically also a (negative) effect on total production and output. It can be debated whether farmers actually act to fully maximise profits on the issue of voluntary set-aside. There seems to be some evidence that farmers prefer cultivating to set-aside, even if this is connected with a slight decrease in income. In order to depict the two possible borderline cases, in the following two sub-scenarios are presented. The two sub-scenarios are named "no ÖPUL" and "no ÖPUL; no premium for voluntary set-aside". The second sub-scenario is identical with the first one, except that there is no premium granted for voluntary set-aside. This scenario can be interpreted as representing the case that farmers prefer cultivation to set-aside and are willing to pay for it. The results of these scenarios without ÖPUL are interpreted against a reference scenario which represents the situation in 1995. The difference between the no ÖPUL scenario and the reference scenario can be interpreted as the effect of ÖPUL.

Table 6.5
Shadow values of land in CSP scenarios in Oberösterreich

	Reference (with ÖPUL)	No ÖPUL		No ÖPUL No premium for voluntary set-aside	
	ATS/ha*	ATS/ha*	% of base	ATS/ha*	% of base
Arable land	5107	3106	61	2772	54
Grassland two+ cuts	4242	2360	56	2360	56
Grassland one cut	3421	2360	69	2360	69
Alpine grassland	2120	1240	59	1240	59

* 100 ATS = 7.27 EURO

6.4.4. Effects on farm income and profits, dual values and budgets

ÖPUL contributes significantly to farm income in the region. Without ÖPUL in place, the total regional gross margin declines by 11% (Table 6.4). Given that the ÖPUL programme is highly differentiated and measures for all types of farms exist, it is plausible that all farm types are affected by this decline. Although a sub-regional distinction of the effects is not directly possible with the aggregated regional model, there are strong indications that the disadvantaged areas within the region profit more from ÖPUL than the favourable regions, both relatively, as well as in absolute terms. Firstly, the relative share of the premiums on total farm revenues (due to lower yield levels) in disadvantaged areas is higher and secondly, the participation rate in the extensification measures, especially in organic farming, is higher in these areas as well. If no ÖPUL would have been in place these regions would have had difficulties to intensify their production.

Very important information can be gained from the dual values of land which are also called shadow prices. These values tell us something about the likelihood that land falls out of production. In technical terms the shadow price gives us the value for which a farmer would rent an additional hectare of land. The lower the shadow price, the more likely it is that land falls out of production and goes idle, which is in Oberösterreich a particular issue for grassland. Actually in the long-run it is likely that lands falls out of production even with a positive though low shadow price in absolute terms. This is because the marginal gross margin which is equal to the shadow price is calculated from the marginal return minus the marginal variable cost—this means fixed (overhead) cost are not included in the calculation, which implies that in absolute terms the shadow prices overestimate the profitability of land use.

Table 6.5 clearly shows that in the reference situation 1995 arable land is much more valuable than grassland and more productive grassland is more valuable than alpine grassland. This is of course to be expected and only demonstrates that the model adequately depicts reality. If ÖPUL is removed from the situation as in the two scenarios presented then the dual values of all types of land are reduced drastically. This observation alone indicates a strong influence of the ÖPUL programme on land value and the upkeep of marginal land. This is especially relevant for alpine grassland whose value drops to a very low level. In the second sub-scenario where no premium for voluntary set-aside is assumed the shadow price of arable land is further decreased. This clearly shows that in the first scenario the shadow price of land is determined by the premium for set-aside.

Two main ÖPUL measures have a significant influence on the shadow values: The 'elementary support scheme' and even more important the 'crop rotation premiums' (the last one only with respect to arable land). Given that both these policies are modelled as direct payments, which

Table 6.6
Structure of direct support to farms in Oberösterreich

	Reference (with ÖPUL)		No ÖPUL		No ÖPUL No premium for voluntary set-aside	
	mn ATS*	%	mn ATS*	% of Reference	mn ATS*	% of Reference
Total premiums	3115	100	1912	61	1916	62
CAP-crop premiums	1061	34[1]	1099	104	1086	102
CAP-livestock premiums	361	12[1]	361	100	361	100
LFA payments	468	15[1]	452	97	468	100
ÖPUL premiums (total)	1225	39[1]	0	0	0	0
Elementary support	348	28[2]	0	0	0	0
Crop rotation scheme	297	24[2]	0	0	0	0
Other ÖPUL schemes	580	47[2]	0	0	0	0

[1] Share on total premium transfers.
[2] Share on total ÖPUL premiums.
* 100 ATS = 7.27 EURO.

do not include farm adjustment, the loss of these transfers cannot be compensated. The reduction of the premium for organic farming or other extensification measures to zero triggers an intensification of production and therefore these losses can partly be compensated by farm adjustment.

The dual value of grassland is reduced by up to 50%, although the reduction in absolute terms is less than for arable land. The dual value is stabilised by LFA-payments, and also the beef extensification premiums stabilise the marginal value of grassland. Without these additional subsidies the dual value is likely to decline to zero.

According to the model results, land abandonment does not take place. This holds for the short run period for which the model was constructed. In the long run however, it cannot be taken for granted that all land is kept in production. If production in such areas is associated with positive external effects they are likely to decline under such a scenario. More precise statements concerning the long term viability of regions with high shares of this type of land are very difficult to give, because of the various income combinations of farmers (share of part-time farmers more than two thirds) and a demand of local tourism industry for countryside stewardship services which may result in a creation of local production premiums financed by local budgets.

The great importance of ÖPUL also becomes obvious if one looks at Table 6.6. ÖPUL makes up almost 40% of the total direct payments to farms in the reference year 1995. Together with LFA (less favoured areas) payments, which account for another 15%, it exceeds the money transferred via general CAP support.

Logically, there is a strong decline in the total transfers if ÖPUL was not in place. The structure and level of the other direct payments is only marginally altered. Both sub-scenarios show an increase in the CAP arable support. It results from a changed cropping structure with higher shares of oilseeds and set-aside (see Table 6.7) for which the premiums are higher in comparison to the standard cereal premium. The LFA payments are slightly reduced in the first sub-scenario again as an effect of the increased set-aside area, for which no LFA payments are granted.

Table 6.7
Cropping structure on arable land in Oberösterreich

	Reference (with ÖPUL)	No ÖPUL		No ÖPUL No premium for voluntary set-aside	
	ha	ha	% of base	ha	% of base
Wheat	51,900	49,690	95.7	51,630	99.5
Grain maize and CCM	36,599	36,816	100.6	37,671	102.9
Other cereals	88,570	77,756	87.8	81,545	92.1
Oilseeds	24,346	24,644	101.2	25,713	105.6
Silage maize	27,057	27,293	100.9	28,053	103.7
Forage crops	33,763	32,440	96.1	33,914	100.4
Set-aside	16,935	29,783	175.9	16,935	100.0

6.4.5. Cropping structure, structure of production methods and output effects

Besides the effect of ÖPUL on agricultural income and the maintenance of agricultural production in marginal areas its effects on aggregate output of agricultural products is of major interest for policy-makers. In the following section the production effects of the ÖPUL programme are analysed in more detail. The final output effects result from three single effects, which are acting simultaneously:

- effect of changes in total cropping area, due to changes in the set-aside acreage
- effects of changed cropping structure (crop rotation)
- intensity effect due to changes in structure of applied production methods (e.g. changes in the fertilisation level)

While in the results of the scenarios presented so far the differences between the two sub scenarios have not been very impressive, with the analysis of the cropping structure (see Table 6.7) and output (see Table 6.9) the distinction between the two sub-scenarios becomes important. This becomes obvious when looking at Table 6.7. If ÖPUL was abolished this would have important side-effects: The set-aside area would increase by nearly 80% in comparison to the reference situation. This is the rather strong reaction that is accompanied by very minor changes in total regional income (compared with Table 6.4). It is immediately obvious that such a change has important implications for aggregate production. This was the reason why the two sub-scenarios were distinguished throughout this section. In the second sub-scenario it is assumed that there is no premium for voluntary set-aside paid. Under these circumstances voluntary set-aside will not be practised by farmers, rather farmers only set land aside which they have to under the EU rules which means that the total amount of set-aside remains unchanged in comparison to the reference situation. We return to the discussion of these effects later when the output effects are analysed.

In addition, beside its impact on the set-aside acreage, ÖPUL has direct influence on the cropping structure. As Table 6.7 shows, ÖPUL promotes the cultivation of more extensive crops, like other (low input) cereals. If there would have been no ÖPUL the acreage of low input cereals would have declined by roughly 10% whereas more intensive crops like maize (grain as well as silage maize) and oilseeds (at first winter rape) would have gained importance.

Table 6.8
Structure of production methods in Oberösterreich

	Reference (with ÖPUL)		No ÖPUL		No ÖPUL No premium for voluntary set-aside	
	ha	per cent	ha	per cent	ha	per cent
Organic and total ban (arable land)	31,010	10.53[1] (100.0)[2]	9,169	3.11 (29.6)	9,722	3.30 (31.4)
Organic and total ban (grassland)	62,813	24.40 (100.0)	54,222	21.06 (86.3)	54,222	21.06 (86.3)
Other extensification (arable land)	111,651	37.91 (100.0)	89,485	30.38 (80.1)	93,140	31.63 (83.4)
Set-aside on arable land	16,935	5.75 (100.0)	29,783	10.11 (175.9)	16,935	5.75 (100.0)

[1] % of respective land category (arable/grassland).
[2] % of reference acreage.

That application of ÖPUL measures leading to a direct extensification of the production on a per hectare base is shown in Table 6.8. In the reference year more than 10% of the arable land is managed according to the rules of organic farming or without mineral fertilisers and synthetic pesticides. In addition, almost 40% of the arable land were cultivated under the restriction of other arable extensification measures Even if all premiums for extensive production would have been reduced to zero, a re-intensification of production is only profitable on some sites, so that still parts of the CSP area are managed in a similar way as with the schemes in place. This is particularly true for those schemes with only relatively "soft" restrictions which therefore lead only to marginal yield effects.

Bearing this in mind it is not astonishing that the area cultivated according to schemes summarised under the heading "other extensification" is only reduced by 20%. The measure supporting the renunciation of growth regulators in cereal cultivation is the most important one in this group. Here it turned out that the application of growth regulators in barley and oats is practically negligible, independently from any premium payment.

On the other hand the acreage of arable land managed according to the guidelines of organic farming and a total ban on agro-chemicals is the most strongly affected type of land use: its area is reduced by two thirds if there is no specific support available. In this context it is important to note that one of the underlying model assumptions is that prices for organic products are the same as those for conventional food. The omission of the premium prices for organic products from the model does not change the results significantly. There are two reasons for this: As only part of the organic products are sold at premium prices (e.g. milk 30–40%, beef 10%, see Michelsen et al., 1999) the behaviour of the marginal producer of organic products in the model is not changed by the omission of premium prices. Also it has to be kept in mind that in the presentation of Table 6.8 organic farming is much less important in terms of land area and quantity of products produced than the technology "total ban of agro-chemicals".

If ÖPUL would not be in place the organically managed grassland would be less influenced than arable land. It would only decline by 15%. This can be explained by the fact that the effects of banning pesticides and mineral fertilisers on yields are considerably weaker on grassland than

Table 6.9
Output effects of CSP-scenarios in Oberösterreich

	Reference (with ÖPUL)	No ÖPUL		No ÖPUL No premium for voluntary set-aside	
	1000 t	1000 t	% of base	1000 t	% of base
Wheat	278	286	102.9	295	105.9
Maize and CCM	346	348	100.7	354	102.4
Other cereals	388	367	94.5	379	97.7
Oilseeds	71	72	101.4	74	104.2
Forage crops + silage maize	197	202	102.6	208	105.7
Cereals consumption in cattle sector	378	362	95.8	353	93.6
Total cereal production	1012	1001	99.0	1028	101.5
Effective cereal supply[1]	634	639	100.8	674	106.3

[1] Total production minus consumption in the cattle sector.

on arable land. The best interpretation of the figure of minus 15% organic grassland is that there is no strong re-intensification on grassland, rather than saying that actually all 85% of this grassland is kept formally under organic farming conditions.

The importance of the set-aside effect for total output becomes obvious if one compares the two different sub-scenarios. In Table 6.9 the total output for different crops is listed. The list is supplemented by some information on the consumption of cereals (feed concentrates) in the cattle sector. The consumption of cereals for cattle feeding is explicitly calculated in order to capture the effects resulting from changes in forage production (and also changes in the size of the cattle herd). Cereals can substitute to some degree other forage products. The difference of total cereal production minus the consumption for cattle feeding is the net supply of cereals which is the most important figure with respect to aggregated output effects.

In the sub-scenario where no premium for voluntary set-aside is granted, the output effects of ÖPUL are pronounced. For all types of crops, except low input cereals, production is significantly higher if extensive production techniques are not subsidised by ÖPUL and the consumption of cereals for cattle feeding is in addition reduced as an effect of an increased production of silage maize. Therefore the net supply of cereals is more than 6% higher without ÖPUL in place, or the other way round, ÖPUL leads to market relief for cereal commodities of the same amount.

In the other sub-scenario, where only ÖPUL premiums are set to zero but the set-aside premium is still available, a lot of arable land is devoted for voluntary set-aside. In principal there is still a tendency to more intensive crops and more intensive production (resulting in higher per hectare yields) leading to a higher production as described for the sub-scenario above if there are no ÖPUL premiums subsidising an extensive production. However, the increase of the set-aside area leads logically directly to a reduction of production, which counterbalances the effect of re-intensification to a large degree. As Table 6.9 shows the total cereal production would be slightly lower (1%) without ÖPUL, but the consumption of feed concentrates would be also lower, so that finally the net supply of cereals is a bit higher (1%) without ÖPUL. The same is true for oilseeds, their total output would also be slightly higher (1.4%) if there is no ÖPUL.

130   *Countryside Stewardship: Farmers, Policies and Markets*

Livestock production is not directly affected by the ÖPUL programme. At a first glance it is not astonishing that there are no changes in livestock production due to the abolition of the ÖPUL programme. Dairy farming and milk production is restricted by production quota and production is so profitable that it is not affected by ÖPUL. Due to a technical relationship the number of calves being produced on dairy farms and used for beef production is also constant. In a longer-term perspective the critical value is therefore the number of suckler cows. Without ÖPUL the income from grassland is in some areas dramatically reduced, due to losses from organic farming premiums and elementary support for grassland. The dual value of grassland declines by 40–50%, but is still positive and no grassland is taken out of production in the short run. Therefore it seems reasonable to conclude that there is no change in the number of suckler cows. This statement is probably not true in a long-term scenario, though. For those livestock activities that are more or less independent from forage production (poultry, eggs, pigs) gross margin is held constant in the model, therefore a constant level of production is technically given.

To summarise, the two calculated sub-scenarios show both the large impact of ÖPUL on farm income, and even more important on the profitability of marginal land. With respect to the quantity of production the development of the set-aside (and in the long-term perspective land abandonment), is of major importance. The first sub-scenario, which represents exactly the CAP rules without the agri-environmental programme, leads to a drastic increase in the set-aside area which implies only small negative effects of ÖPUL on production. However, there are some indications that farmers at least in a short term perspective might not set-aside land voluntarily in the same amount as calculated. The first point is a methodological one and linked with the aggregation error immanent in the feeding optimisation process. The second argument refers to the general behaviour of farmers. There are some indications that farmers prefer to manage and cultivate their land, and have some reservations with respect to set-aside farm land, at least as long as the financial consequences of this behaviour are minor, which is true in our case. Although it is very difficult to quantify the actual effects of these two topics, it is quite probable that in a short term perspective voluntary set-aside might only be applied to a more limited extent than pure profit maximising behaviour might suggest. This is supported by the small difference in the total regional gross margins (see Table 6.2) of the two sub-scenarios compared. So finally following conclusions can be drawn:

- In a short-term perspective ÖPUL leads to a significant reduction in total output of roughly 6% with respect to the main arable crops, whereas the livestock sector is not directly affected.
- In the more long-term perspective the predominant effect of ÖPUL is the avoidance of land abandonment. Agricultural land is kept in production, which counterbalances largely the relieving effects on commodity markets resulting from the application of more extensive production methods offered by ÖPUL.

6.5. The effects of CSPs in the other study regions

In general the CSPs analysed, lead only to limited market and production effects. However significant regional differences can be observed in the different case studies. For Germany (Oberpfalz, KULAP programme) and UK (Cumbria, ESA) the results indicate that the CSPs lead here to a limited but unambiguous reduction of output. In the Oberpfalz the net cereal supply is reduced by roughly 2.5%, whereas in Cumbria the ESA restrictions effect mainly the breeding ewe herd, which is reduced by about 5%.

For the other study regions the production effects are more or less negligible with respect to the commodity markets under consideration. It is a quite simple observation, that the aggregated production effects result from two factors, which are connected by multiplication. The first factor is the output effect per hectare, the second one the area under policy or the participation rate respectively. Both factors are relevant to explain the small overall production effects.

As one can see from Table 6.1, a lot of CSP schemes focus on income support and upkeep of agricultural land use (*Prime à l'herbe* (FR), elementary support (DE, AT), Swedish schemes), with practically negligible output effects on a per hectare base. In other regions the CSPs aim mainly at commodities, which are not regarded here in detail, like permanent crops (Veneto, Italy) or cotton (Thessalia, Greece).

The effects of organic farming on total output are very significant on a per hectare base. In the model calculations these are only captured as long as organic farming is mainly motivated and determined by direct payments (as in the case in Austria). If conversion to organic farming is chiefly determined by market influence (premium prices for organic products), the production effects of organic farming are not counted as a CSP effect. Thus organic farming has significant output effects, but the acceptance of organic farming and thus the market effects might not always be triggered by policy influence.

Budgetary limitations are one reason for low participation rates. They lead either to a direct refusal of applications or result in an unattractive policy design (premiums too low). The design of the CSPs and also of the general CAP has further implications for the participation rate.

To obtain a measurable output reduction it is necessary that participation is high in such measures that have direct negative effects on yields. It seems that the necessary participation can easier be achieved with relatively "soft" restrictions (such as a ban on fungicides or growth regulators) than with strong ones (a total ban of agro-chemicals), because the required production technique does not differ very much from the production technique farmers are used to. Such a more gradual approach, however, requires a strong regional differentiation of restrictions as well as of direct payments, to ensure the effectiveness of the schemes and to avoid exaggerated over-compensation. Such a differentiation as well as higher payments might be necessary if one wants to attract farmers on favourable sites into CSP participation. The higher degree of differentiation might in turn lead to increased transaction and control cost.

From the model calculations it seems to be obvious that it would be desirable to grant general CAP support in a more decoupled way, especially with respect to the livestock sector. This would ensure that CSPs do not have to work (via high direct payments) against CAP support that is coupled with the production quantity. A general CAP support coupled with product quantity increases the opportunity for yield losses or reduced quantities caused by CSPs, and is therefore an impediment to a higher uptake of CSPs. This is clearly shown by the Cumbria case study. A further indication of this is that in general the participation in CSPs which require a direct reduction of livestock is very limited. Also the model results show in almost no case (except Cumbria), a reduction of livestock due to CSP participation. Effects of CSP on forage production are always compensated by substitution with other feeding components. The proposals for the arable sector of Agenda 2000 lead therefore in the right direction.

CSPs of course do not only have an effect on production but also on farm income and profitability of the use of marginal land and thus on land abandonment. With respect to income support and the avoidance of land abandonment the CSPs analysed seem to be effective instruments, at least in comparison with other CAP instruments. The share of income support as a proportion of the total CSP payment is higher on less favourable sites. The stabilising effects of CSPs on the dual values (marginal value) of different land categories, is always higher than the average effect on the regional gross margin. Apart from the effect of keeping agricultural land in use, none of the CSPs analysed gives an incentive to increased production by intensifying agricultural practices.

In summarising it can be stated that among the numerous CSPs analysed within this task, two main groups of scheme have been identified.

The first group of measures addresses mainly the **maintenance of agricultural land use**. This group consists of the French *Prime à l'herbe* scheme, the elementary support schemes in Austria and Bavaria (Germany) and most of the Swedish schemes. No major market effects, in a sense that output is reduced, can be expected from these schemes. They mainly stabilise farm income and (shadow) prices of land. However these policies are nevertheless quite different from general CAP support instruments. In contrast to the general CAP instruments these CSP schemes are more focused, that means they exclude often intensive farms on favourable sites. A second criterion is, that the schemes do not give an incentive to intensify production in any case.

The second group of schemes consists of measures which lead actually to an **extensification** of production, combined with decreasing yields or stocking densities per hectare. From a broad application of these types of schemes measurable market effects can be expected. Often they are of a horizontal design with uniform premium payments, so that they also include an income support element. The less favourable the site on which farmers apply these measures the higher is the share of income support.

The main effect of the CSPs analysed is to keep agricultural land in production and to avoid land abandonment. This is partly combined with a reduction of negative externalities from agriculture and with limited output reduction effects on a per hectare base. If one accepts the avoidance of land abandonment being a major goal of agricultural policy, the CSPs investigated perform better than other general CAP support instruments because they are more focused to the problem.

6.6. CAP, Agenda 2000 and their relevance for CSPs

Finally we discuss some of the implications of the CAP and Agenda 2000 for CSPs. Although the calculations for Agenda 2000 were based on the proposals of the Commission of March 1998, the general statements derived form them are still valid for the most recent version of Agenda 2000 decided upon at the EU conference in Berlin (March 1999), because the principal direction of the policy changes (lower guarantee prices, increasing direct payments) stay unchanged. However, the discussion is limited to the qualitative insights gained from the quantitative model calculations.

Although part of the general CAP rules, the **beef extensification scheme** could be classified as a CSP. This is the case for two possible reasons:

1. The beef extensification scheme gives an incentive to reduce production
2. The scheme contributes to the objective of avoiding land abandonment

The modelling results show a mixed picture with respect to both points. The output effect of the scheme is rather ambiguous. An output reduction due to the extensification premium can only be expected if the "standard" livestock density (derived from the yield potential of the forage area) is slightly above the limit given by scheme. In this case the scheme might offer an incentive to reduce stocking density in order to obtain the premium. If the "standard" livestock density is much above the one of the scheme, the payment is not sufficient to induce participation. If the "standard" livestock density is clearly below the limit, the scheme might lead to an intensification, because the premium is paid on a per head and not on a per hectare basis.

With respect to the objective of avoiding land abandonment it is obvious that any income transfer will contribute to keeping agricultural land in production. However, the effectiveness is more limited the lower the standard stocking density and the more heterogeneous the sites are within a farm.

In order to avoid incentives to increase production a substitution of the beef extensification premium by a regionally differentiated grassland premium paid on a per hectare basis might be useful. The premium level should be based on the yield potential of the grassland and it should also be higher for meadows in comparison with pastures.

If Agenda 2000 was accompanied by the existing CSPs participation rates would increase. This is quite logical because the opportunity cost of yield losses due to CSP are valued lower under the lower commodity prices coming with Agenda 2000. However, this is probably a rather minor effect. If it is intended that CSP have major output effects it would be necessary to review and reshape the policies in a more fundamental way. Agenda 2000 potentially has effects on less favourable areas which are important in the context of CSP. In the arable sector the proposed price reduction is not fully compensated for by higher per hectare payments. However, the effects on marginal areas depend mainly from the degree of regional differentiation of the premiums. Marginal areas will tend to benefit from a limited degree of differentiation.

Although grassland is not directly addressed by Agenda 2000, the policy plan nevertheless has significant impact on grassland profitability and competitiveness, especially in a more long-term perspective. Due to reduced opportunity costs of arable land and feeding concentrates, which finally determine the value of grassland, the latter will loose competitiveness, because there is no direct compensation for these losses. The limited effectiveness of the beef extensification scheme with respect to the profitability of grassland use has already been discussed above. The effects of the Agenda on grassland are reinforced by technical and biological progress leading to increased milk yields per cow, and therefore in connection with the quota (respectively limited demand for milk and its products) to a decreasing number of cows, so that finally grassland becomes more and more abundant.

Thus a somewhat brighter perspective on marginal arable land seems more likely than on marginal grassland. While the arable land will be less likely to be abandoned, the contrary is probably true for grassland. This indicates a need to place special emphasis on grassland in the upcoming agri-environmental regulation (following Regulation 2078/92) and its regional implementation.

References

Bauer, S. and Henrichsmeyer, W. (1989) *Agricultural Sector Modelling*. Wissenschaftsverlag Vauk, Kiel.

Dabbert, S., Herrmann, S., Kaule, G. and Sommer, M. (1999) *Landschaftsmodellierung für die Umweltplanung. Methodik, Amwendung und Übertragbarkeit am Beispiel von Agrarlandschaften*. Springer, Berlin.

Day, R. H. (1963) On Aggregating Linear Programming Models of Production, *Journal of Farm Economics*, 45(4), 797–813.

Deblitz, C. (1998) *Vergleichende Analyse der Ausgestaltung und Inanspruchnahme der Programme zur Umsetzung der VO (EWG) 2078/92 in ausgewaehlten Mitgliedsstaaten der EU*. Landbauforschung Völkenrode Sonderheft 195, Institut für Betriebswirtschaft, Agrarstruktur und ländliche Räume, Bundesforschungsanstalt für Landwirtschaft (FAL) Braunschweig.

Deblitz, C. and Plankl, R. (Eds) (1998) *EU-wide Synopsis of Measures according to Regulation REG (EEC) 2078/92 in the EU*. Institut für Betriebswirtschaft, Institut fur Strukturforschung, Bundesforschungsanstalt für Landwirtschaft (FAL) Braunschweig.

Frede, H-G. and Dabbert, S. (Eds) (1999) *Gewässerschutz in der Landwirtschaft. 2nd Edn*. Ecomed, Landberg, Germany.

Howitt, R. (1995) Positive Mathematical Programming, *American Journal of Agricultural Economics*, 77 (May), 329–342.

Michelsen, M., Hamm, U., Wynen, E. and Roth, E. (1999) *The European market for organic products: growth and development. Organic farming in Europe: economics and policy, Vol. 7*. Universität Hohenheim, Stuttgart-Hohenheim.

Zeddies, J. and Doluschitz, R. (1996) *Marktentlastungs- und Kulturlanschaftsausgleich (MEKA). Wissenschaftliche Begleituntersuchung zu Durchführung und Auswirkungen*. Ulmer, Stuttgart.

G. Van Huylenbroeck and M. Whitby
Countryside Stewardship: Farmers, Policies and Markets
© 1999 Elsevier Science Ltd. All rights reserved

Chapter 7

Estimating aggregate output effects

Franz Sinabell, Klaus Salhofer and Markus F. Hofreither

Abstract—Most of the agri-environmental policies aim at environmental goals by restricting inputs and offering direct payments to farmers in order to motivate them to participate. In the following sections several findings from other chapters of this book are combined with an equilibrium displacement model to show that countryside stewardship policies are not a set of policies from which immediate output reductions can be expected. It is shown that two counteracting effects exist: an output decrease due to input restrictions, as well as an output increase due to direct payments. Hence, the overall effect on output at the market level is ambiguous and critically depends on how farmers use the premiums they get when they enrol in a particular countryside stewardship scheme. Responses from the survey of farmers' attitudes towards agri-environmental programmes (see Chapter 5) confirm such ambiguity because only a minority of farmers reports output reductions as a consequence of programme participation. The equilibrium displacement model is applied empirically to the most important EU countryside stewardship policy aiming at input restriction (the Austrian 'crop rotation scheme') and the results show that in the short run hardly any output effects can be observed if farmers use part of the direct payments to finance additional non-restricted inputs.

7.1. Introduction

During the seventies it became evident that modern agricultural practices do have substantial impact on the natural environment. Negative external effects due to intensive farming practices occur such as erosion, soil compaction, and water pollution. In addition, public awareness of these negative externalities as well as of positive external effects like landscape amenities which are 'by-products' of certain farming systems increased (see Baldock and Lowe, 1996). To cope with negative externalities, environmental legislation has been introduced which has frequently been integrated in agricultural policy regulations. Typical examples of such an approach are cross-compliance policies or limitations of livestock densities. The provision of positive agricultural externalities for a long period was thought to be best addressed by promoting agricultural production via price support policies.

During the GATT negotiations on agriculture the European Community was exposed to considerable pressure exercised by its trading partners to reduce the level of support given to farmers and to abandon trade distorting agricultural policies. The Community reacted by launching a fundamental reform of the Common Agricultural Policy (CAP) in 1992. The central element of this reform was the shift from price support to direct payments, aiming to combine control of agricultural markets with extensification of agricultural production.

The policy shift from price support policies to direct payments follows the rationale to de-couple support to farmers from output. De-coupled policies are viewed not to be trade distorting and are part of the so-called 'blue box' or 'green box' of support policies. The level of support from such policies needs not to be reduced as is the case with many other policies. Environmental programmes are explicitly part of the 'green box' if, among other criteria, premiums are not higher than the additional cost necessary to reach a given environmental goal. These and other aspects relevant to international trade are treated in detail in the next chapter.

As part of the so called accompanying-measures of the CAP reform, Council Regulation 2078/92 was established which is designed in a way to be consistent with the definition of 'green box' policies. According to the reasoning of this regulation the role of farmers is not seen to be just producers of agricultural commodities, but also to be stewards of the environment and the countryside.

Based on Regulation 2078/92 Community aid programmes have been introduced in all EU Member States. The objectives of these programmes that are part-financed by the EAGGF (European Guidance and Guarantee Fund) are:

- to accompany the changes to be introduced under the market organisation rules,
- to contribute to the achievement of the Community's policy objectives regarding agriculture and the environment,
- to contribute to the provision of an appropriate income for farmers.

Clearly, the goals of this regulation are to reduce farm output and/or curtail environmentally detrimental side effects of farm production while supporting farm income.

However, Member States do have considerable freedom to put more or less weight on each of the above policy goals, to choose among the particular targets, the instruments to reach them, and the amount of funds they deem appropriate to attract farmers to enrol. Any such programme may consist of several schemes which are offered either in the whole country or in particular regions only. Two major groups of such programmes can be differentiated according to their major objective: reduction of negative external effects, and promotion of the upkeep of agricultural production in marginal regions. In the remainder of this paper the heterogeneous set of aid programmes emanating from Regulation 2078/92 will be subsumed under the term countryside stewardship policies (CSPs).

Since 1993 a total of 163 programmes has been notified for adoption by the Commission and 152 programmes had been adopted by the end of 1996 (Scheele, 1996). The volume of premiums paid to farmers over the same period is totalling 6.9 billion ECU (Deblitz, 1998). Total budgets of this programme are distributed mainly between Germany (24%), Austria (21%), France (13%), Italy (11%), and Finland (11%). The share of agricultural land which is managed according to at least one of the programme requirements is 87% in Austria, 78% in Finland, 76% in Luxembourg, 45% in Sweden, 30% in Germany, 20% in France and below 10% in the other EU Member States.

The most important of these CSPs—as far as acreage and transfers are concerned—are aiming at the introduction or maintenance of extensive production in marginal areas. Thus by definition hardly any output changes, compared to the pre-policy situation, can be expected due to such policies in the short run. However, such changes are expected as a consequence of the other group of policies—those using input restrictions which are the most frequent ones. The first of the three goals mentioned above, namely output reduction, will be investigated in more detail in this chapter.

In most cases CSPs attempt an extensification by utilising standard instruments, like input control, and compensate farmers via direct payments. While the direct impact of these policies clearly is a reduction of output, the direct payments may create additional input demand and hence weaken or even reverse the output effect of CSPs. Both effects can be evaluated using an equilibrium displacement model (EDM), a tool frequently used in agricultural policy analysis (e.g. Gardner, 1987; Alston, Norton and Pardey, 1995, Chapter 4; OECD, 1995).

EDMs can be used in both, *ex-ante* as well as *ex-post* analyses and allows clear identification of the aggregate effects of a single policy. This chapter therefore offers a complementary approach to the regional partial equilibrium models presented in the previous chapter in which the simultaneous effects of a set of policies are investigated. Data obtained from the farm survey which is treated in more depth in Chapter 5 are used for the empirical specification of the EDM to offer an explanation for the fact that many participants of CSPs reported that they did not experience significant output reductions.

In the following section CSPs are classified according to the instruments that are used and the role of direct payments is discussed in the context of property rights in Section 7.3. Methodological approaches to evaluate output effects are discussed in Section 7.4. In Sections 7.5 and 7.6 empirical findings are presented. For illustration purposes the Austrian 'crop rotation scheme' was chosen because it is among the most important CSPs using input restrictions in the EU and representative of the class of policies which is most frequently implemented. The results of this analysis are discussed in the last section. Conclusions are drawn as to how effective the CSPs analysed are with respect to reducing output, one of the goals of Regulation 2078/92, and how the results of this model can be interpreted in the context of findings on the market effects obtained from other methodological approaches.

7.2. Classification of CSPs according to the instruments that are used

In Chapter 2 an inventory of CSPs is presented which covered both, policies emanating from Regulation 2078/92 and others which are based on national initiatives or legislation (summarised in Annexe). In another inventory, edited by Deblitz and Plankl (1998), details of over 800 Regulation 2078/92 schemes in all the EU Member States except Luxembourg are characterised. As can be concluded from both inventories there is a great diversity in almost all aspects ranging from objectives, transfer vehicles, premiums, and the acceptance by farmers. Virtually the only feature Regulation 2078/92 measures have in common is that they are voluntary, leaving it to the single farmer to participate or not. In Chapter 3 a thorough treatment of features of agri-environmental programmes is presented and groups of similar policies are identified. For the purpose of this analysis a simpler approach is adopted by concentrating on the general mechanisms of such policies and on the instruments they use.

The general mechanisms of Regulation 2078/92 measures are:

1. Society leases property rights of farmers for some period (at least five, in some schemes up to 20 years) and compensates farmers usually on a per hectare basis mostly depending on the opportunity cost of standard land-use.
2. Society sets up quasi-markets for countryside stewardship goods and buys services that have the character of public goods. Usually payments are not delivered on a per unit of good basis.

Typical of category 1 are schemes aiming at reducing negative impacts of farming methods (like erosion, emission of minerals and other farm chemicals) and extensifying farm production (like reduction of the proportion of sheep and cattle per forage area). Schemes falling into category 2 are less important with respect to both their number as well as the volume of transfers they generate. Under such schemes farmers manage their land to preserve habitats and enhance biological diversity. Both mechanisms are at work in schemes where farmers allow access to their land for recreational purposes and provide management and infrastructure.

The set of instruments on which CSPs in the EU are building are almost entirely among the classical instruments of agricultural policy:

(a) output control;
(b) input control;
(c) production premiums; and
(d) various combinations of these instruments.

Pure output control measures generally do not play an important role in CSP programmes. However, in many of the schemes the number of livestock per hectare of land is restricted to a maximum number. These restrict output only in cases where participating farmers cannot buy or lease additional land to get below the respective threshold value.

Input control measures, on the contrary, are of major importance. They frequently restrict the use of:

- land as a factor of production: set-aside schemes (mostly motivated by water protection concerns or habitat development) and schemes aiming at converting arable land into grassland;
- purchased inputs: schemes limiting the use of mineral fertiliser and/or pesticides or banning them almost entirely like in the 'organic farming scheme' which is offered in all EU Member States.

In a number of cases the use of special equipment is requested from participants, e.g. in order to reduce soil erosion. By prohibiting the use of standard machinery such schemes fall into the class of input restrictions (in this case durable equipment) with the effect of increasing production cost.

Production premiums are paid in schemes aiming at increasing the number of head of local livestock breeds and schemes trying to motivate farmers to plant crops in danger of extinction. In addition there are schemes with premiums not directly linked to output but having rather similar effects, e.g. by paying premiums per head of livestock grazing on marginal land. The latter schemes are intended to prevent land abandonment and thus contribute to the maintenance of the provision of landscape amenities that are coupled products of certain extensive farming systems.

A typical example of combinations of both, output and input control measures, is the Austrian 'elementary support scheme' (limiting the number of livestock per hectare land and restricting fertiliser use). Among rather complex schemes, combining output and input control with production premiums is the French *'Prime à l'herbe'* scheme. Premiums are paid if stocking rates lie between specified minimum and maximum boundaries while simultaneously the use of farm chemicals is restricted. Many of the schemes in the United Kingdom add further complexity by combining these instruments with requirements to open land for public access with the associated requirements to provide infrastructure and management.

The effect of programmes aiming at the upkeep of agricultural production are obvious, the premiums paid in such schemes ensure the maintenance of production. Usually the stocking rates in such schemes have to be between a minimum and maximum range to prevent overgrazing and abandonment of pastures and therefore output is constrained within a given range. Due to the design of such programmes output effects, compared to the pre-policy situation, cannot be expected in the short run. In fact, the largest Regulation 2078/92 programmes, in terms of area covered, fall into this category.

Data of the inventory presented in Chapter 2 are used to give an overview of how many measures are using restrictions on inputs to achieve environmental objectives (see Table 7.1). For more detail, see the Annexe of this book where the inventoried policies are classified according to the instruments used. Only Regulation 2078/92 measures are chosen for the figures in Table 7.1.

Table 7.1
Percentage of Regulation 2078/92 measures in eight EU Member States using input restrictions or restricting livestock densities or the use of manure and farm wastes

	Ban/set-aside		Reduction/limitation/other restriction	
	% of schemes	% of funds	% of schemes	% of funds
Herbicides/pesticides	42	8	18	7
Mineral fertiliser	35	5	16	11
Water	6	2	5	1
Manure/farm wastes	12	10	36	9
Livestock	–	–	30	21
Land	5	<1	25	n.a.

Source: STEWPOL CSP inventory data, Gatto (1999); Deblitz and Plankl (1998); own calculations.
Notes: For 118 of 225 programmes financial data were available. In case financial data were available for more than one year for a given scheme, the year with the highest amount was chosen.

Most of the schemes either ban or restrict the use of pesticides and (mineral) fertiliser. In 5% of Regulation 2078/92 schemes farmers are required to set-aside land for 20 years. According to the data in the inventory edited by Deblitz and Plankl (1998) in 25% of the cases either arable land must be converted to grassland/pasture, or the conversion of grassland/pasture to arable land is prohibited, or some other restriction on land is implemented. How such restrictions work at the market level is demonstrated in Section 7.6 and empirical results of an important and representative scheme will be presented there.

In Table 7.1 organic farming is classified to ban mineral fertilisers, pesticides and herbicides (thus some of the classifications of CSPs from the Annexe were used in a slightly modified way). There are other schemes in which farmers can either choose between the 'ban' or 'reduction' option—such schemes are counted in both categories. Since many of the programmes are multi-objective in addressing more than one of the items listed in Table 7.1 figures do not sum to 100%. The percentages rather can be interpreted as a frequency list of instrument categories.

To avoid multiple counting with respect to funds, premiums are equally distributed over policies which restrict more than one item (e.g. funds for organic farming are equally distributed over 'ban mineral fertiliser' and 'ban herbicides/pesticides') with one exception: premiums for setting aside land are not distributed over other categories because of the substitution relationship between land and the other inputs.

This procedure gives a crude estimate of how Regulation 2078/92 premiums in eight EU Member States were spent depending on which instruments were used. Most of the money is spent for schemes in which input restrictions (mineral fertiliser, herbicides, pesticides, water, and land) are used. Funds spent for them account for 34% of all Regulation 2078/92 premiums according to the calculation procedure outlined above. Premiums for measures restricting livestock density which are frequently used in programmes aiming at avoiding land abandonment account for 21%. The remainder is used for restricting the use of manure or farm wastes (19%) or for purposes not listed here (25%).[1]

---

[1] Since 'reduction or limitation of livestock' and 'restrictions on farm wastes and manure' could be seen to belong to one group instead of two, an alternative calculation was made as well: if they are treated as an aggregate, this group accounts for 37% of the funds which is then equal to the funds for the aggregate of the input restricting programmes (not shown in Table 7.1). This decrease from 40% (10% + 9% + 21%) to 37% and the increase from 34% to 37% for input restricting programmes is brought about by different procedures to allocate funds to these categories.

7.3. The role of property rights and direct payments

Countryside stewardship policies—as characterised above—comprise generic environmental legislation focusing on the prevention of negative external effects of agricultural production. Among the most important environmental concerns is water pollution and many policies in the EU Member States either ban or limit the use of farm chemicals or aim at reducing mineral emission by restricting livestock production. Such policies legally oblige farmers to comply with these regulations and therefore are defining how property rights are allocated between agricultural producers and the public. If, for example, a water act states that stocking rates may not be higher than a certain number of livestock per hectare farm land, farmers who reduce their livestock below this limit do not fully make use of their property rights. The reasons for not fully utilising the property rights may be that the optimal production plan requires a more extensive production or that the farmer is concerned about the environmental consequences of mineral emission and sacrifices profits because of his/her environmental attitudes (Chapter 5; Vogel, 1998). However, if farmers do not reduce stocking rates voluntarily, the only possibility of motivating a producer to adopt a more environmentally friendly way of production (reduce stocking rates) is either by compensating him or her for the profits foregone and/or by trying to influence environmental attitudes.

Many schemes of Regulation 2078/92 programmes are designed to work in such a way. These programmes motivate farmers to make further efforts to go beyond minimum standards that are required by compulsory agri-environmental regulation. In addition, a smaller set of programmes focuses on the upkeep of production in marginal areas and therefore aims at preventing land abandonment or afforestation of agricultural land. In order to attract farmers to participate in agri-environmental programmes compensatory payments are paid which can be interpreted as the price for the temporary transfer of property rights from the farmer to the public. Such payments are rationalised in Article 5, 1b of Regulation 2078/92 by the need to at least compensate the loss of income and in addition provide an incentive to make such schemes attractive.

Figure 7.1 depicts a situation in which the public offers a certain price for an environmental good or service. If we assume a horizontal demand (uniform premium) and a normal short run supply schedule S it is obvious that only for the marginal participant of an agri-environmental programme

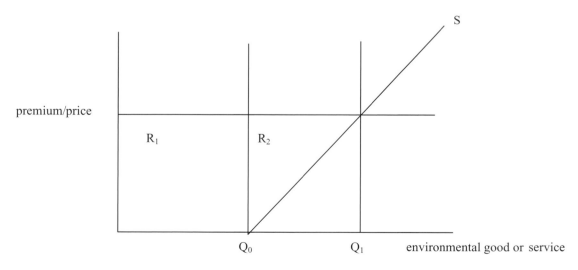

Fig. 7.1. **A market of environmental goods and services which are coupled products of agricultural production.**

will costs (or profits foregone) equal premiums obtained. If in the reference situation the quantity Q_0 of a particular environmental good is produced as a by-product of agricultural production the public can stimulate an increase in production to output Q_1 by offering a premium (see horizontal demand schedule). The rectangle between Q_1, the demand function, and the axes represents total transfers to farmers and rectangle R_1 and triangle R_2 are equivalent to rents accruing to participants if we assume that cost (including opportunity cost) are below short run supply schedule S.

A way to reduce the total volume of premiums and consequently the rents is to buy just the quantity between Q_0 and Q_1. This is frequently done in Regulation 2078/92 programmes by offering a particular scheme only in confined sub-regions where local scarcity of a particular environmental good has been observed (e.g. in Sweden only producers in designated areas where forestry is the dominant land-use are motivated to provide the good 'open landscape'). An alternative mechanism which would even eliminate the triangle R_2 would be to make an auction and accept bids from farmers until the optimal quantity is reached. Such mechanisms are in general not used in Regulation 2078/92 programmes. Policy mechanisms of agri-environmental programmes are treated in more depth in Hofreither (1998), Latacz-Lohmann (1998), and Weaver (1998).

Another aspect of Regulation 2078/92 which may raise equity concerns can also be shown using fig. 7.1. It may be the case that in country A the reference level of 'good agricultural practices' is defined in a way that requires farmers to provide Q_1 of an environmental good and the property right lies with the public. This can, for example, be the case if meadows may not be mown during a specified time period in order to protect a certain bird population. In country B the same species may not be protected at all and therefore a voluntary programme may be the only tool to provide such protection apart from changing nature conservation legislation. In country A farmers are legally obliged to provide this service and incur the cost below the supply schedule while in country B participants of a biodiversity programme can obtain the rents shown in fig. 7.1. This model can help to explain why in different countries programmes with compensatory payments having similar goals are implemented quite differently while in other countries they are not offered at all. Other explanations for differences in cross-country comparisons may include the demand for environmental goods and services and/or other policy preferences concerning the spending of public funds.

Besides such implications agri-environmental compensatory payments can also be seen as inefficient because it may be the case that farmers need not to change their behaviour to obtain premiums. This will particularly be true if horizontal measures are applied uniformly across a whole country.

Moreover, the question arises how premiums are used on the farm. Clearly, depending on the mechanism chosen, many or most farmers will use the premiums to finance the additional effort to carry out the required stewardship practice, but it is not certain that all the remainder is used for consumption purposes by the farm household. Such payments could be used to subsidise non-restricted inputs or to finance investments and therefore can have consequences at the market level. The implications of these considerations will be analysed in detail in Section 7.5.

7.4. Methods to evaluate market effects

Several approaches to evaluate the market effects of policies can be identified including:

1. methods using farm models and aggregating the results to the market level;

2. the use of aggregate demand and supply models and specifying them according to market information;
3. econometric methods can be used to measure the impact of CSPs by analysing time-series data;
4. asking farmers how they reacted to policy changes and drawing conclusions about the aggregate effects;
5. a qualitative way to learn more about the likely effects of policies can be made by benchmarking them against a set of criteria.

Methods falling in category 1 frequently use linear programming models. The advantage of such models is that comparably few data are needed to obtain results on an array of different farm types and that ex-ante simulations are possible. A limitation of such models is that usually many assumptions have to be made concerning technical coefficients. Another shortcoming of such models is a certain lack of transparency and difficulties in aggregating the results to the market level.

Aggregate demand and supply models like the one presented in Section 7.6 depend on assumptions about the behaviour of agents in a market situation and can be used in both, ex-post as well as ex-ante situations. In an ex-ante analysis assumptions have to be made about key parameters and the results show in which direction variables are likely to be moving in a policy experiment. When market data are available, in addition ex-post analysis can be made and the actual consequences of a policy can be analysed. The advantage of such models is that they are comparably transparent and can be developed quickly. As is often the case with other empirical models, finding the necessary data to specify them empirically is not always possible and therefore sometimes assumptions about the values of parameters have to be made.

A method that combines advantages of both, category 1 and 2 approaches, is the method of positive mathematical programming (see Howitt, 1995; an application is provided in Chapter 6). With this method market level results can be obtained that are based on observed data and thus a major disadvantage of LP-models can be overcome. As in LP models, policies can be analysed in a very detailed manner. The disadvantages of this method are that in some cases aggregation errors can occur and that this method is limited to ex-post analysis.

Category 3 models require a considerable amount of data, which are hardly available for many CSP schemes since they have only been introduced recently. Moreover, since only a small share of EAGGF transfers is used for CSPs, significant coefficients of a policy variable in an econometric model are unlikely to be identified in particular if output effects are small.

Using surveys to obtain information on the output effect of a particular policy often is the only choice available if parameters necessary to specify economic models cannot be obtained from market data. Since agri-environmental schemes are frequently tailored to specific regions for which data cannot be found in national statistics, surveys are the only method to obtain data. However, even if great care is taken to ask the right questions and avoid interview bias, the results have to be interpreted with care. The reason is that in many cases farmers do not have records of the relevant data and the responses therefore must be based on judgement rather than hard facts. A disadvantage of the survey method is that representative results can only be obtained at very high cost.

Since many of the Regulation 2078/92 programmes have been introduced only recently, missing data limit the use of quantitative tools. The empirical results presented in Sections 7.5 and 7.6 give some indication of the overall market effects but the number of policies that can be analysed in a quantitative way is limited. A category 5 analysis as presented in Chapter 8 of this book can provide evidence whether CSPs are likely to be market distorting due to their design.

In the next section farm survey data are used to draw conclusions about the likely market effects of CSP programmes and these findings are discussed in the context of the results from a quantitative aggregate demand/supply model in Section 7.6.

Table 7.2
Participants of CSPs who have reduced arable land or livestock and have purchased new machinery as a consequence of programme participation

EU Member State	Programme participants have		
	Reduced arable land (%)	Reduced livestock (%)	Purchased machinery (%)
Austria	14	22	24
Belgium	4	10	22
France	13	6	13
Germany	35	23	22
Greece	4	2	17
Italy	7	8	16
Sweden	13	16	17

Source: STEWPOL farm survey data, Bergström, 1999.
Notes: $N = 924$ (arable land); $N = 999$ (livestock); $N = 1008$ (purchased machinery). Inside-country-weights were applied to account for the fact that stratified samples were used

7.5. Results on the output consequences of CSPs from a farm survey

The 1998 farm survey interviewed nearly 2000 farmers, asking about their attitudes towards agri-environmental programmes (see Chapter 5). This survey included several questions relating to output effects of CSPs. The most important information obtained from the farm interviews for this study are the responses of farmers on:

- how does farm output change when farmers enrol in a CSP scheme?
- how do farmers spend the money they receive as a compensation for complying with the requirements of CSP schemes.

Information on the second question is an essential input to the empirical model that will be presented in the next section.

The figures in Table 7.2 show how many participants of agri-environmental programmes have reduced the share of land allocated for arable crops, reduced the number of livestock and how many among them needed new machinery. The figures show major differences among the countries with respect to reductions of arable land and heads of livestock. What is surprising is that in all countries more than 10% of the participants needed new machinery. To summarise the results, less than one third of participants reduced the share of land for arable crops and less than a quarter reduced the number of livestock.

Participants of agri-environmental programmes were asked about the long term output consequences on their farm (See Table 7.3, middle column). Only in Germany and Austria more than 50% of the participants expressed the view that total farm output is decreasing. Some farmers reported that output actually is increasing and not decreasing as would be expected from the fact that the majority of CSPs is limiting inputs.

Non-participants were asked to identify the two major reasons why they did not participate. From the whole sample 18% of non-participants ranked negative consequences on farm output among the two most important reasons for not participating (the country figures are reported in Table 7.3, right column). Other major reasons for not participating are that premiums are

Table 7.3
Share of participants and non-participants view on the long term output effects of CSPs on their farm

EU Member State	Participants observe	Non-participants are concerned
	Output decreases (%)	About negative output effects (%)
Austria	56	0
Belgium	35	11
France	18	27
Germany	51	33
Greece	19	36
Italy	15	5
Sweden	18	5
United Kingdom	38	15

Source: STEWPOL farm survey data, Bergström, 1999.
Notes: $N = 1170$ (participants), $N = 810$ (non-participants). Inside-country-weights were applied to account for the fact that stratified samples were used. The question for the participants was: 'What will be the likely long-term impact of the schemes on overall farm output, compared to a situation where such schemes did not exist? Think of your own farm!' The question for non-participants was 'What are the two main reasons why you have not joined any management agreement scheme?' Among twelve options respondents could choose: 'I think the schemes will have a negative impact on my farm output.' N.B. only 5 out of 250 farmers from Austria did not participate in any scheme.

too low (25%), that respondents felt not well enough informed about the agri-environmental programmes (22%) and that the application process is too burdensome (8%). Several more questions related to output consequences were asked in the survey. The answers that were given are consistent.

The results obtained from the farm survey with respect to the consequences of CSPs on farm output show that only in a minority of cases output is actually declining when data are aggregated over countries. This might be explained by the fact that only some farmers participated in schemes which require input reduction while others participate in schemes aiming at keeping marginal land in production which in general maintain production without implementing a binding constraint on input use. Another explanation which is treated in more depth in the following sections could be that at least some farmers use part of the direct payments to buy additional, non-restricted inputs, which outweighs the output decrease of input restrictions.

7.6. Quantitative output effects of countryside stewardship schemes

7.6.1. The method of equilibrium displacement models

An expedient way to analyse CSPs is the 'equilibrium displacement model' (EDM) approach in the tradition of Muth (1964), Floyd (1965), and Gardner (1987). The EDM chosen for this study represents one output market for the agricultural product as well as two input markets, one which is directly affected by the CSP, e.g. through an input restriction, and one which represents all other inputs as illustrated in fig. 7.2.

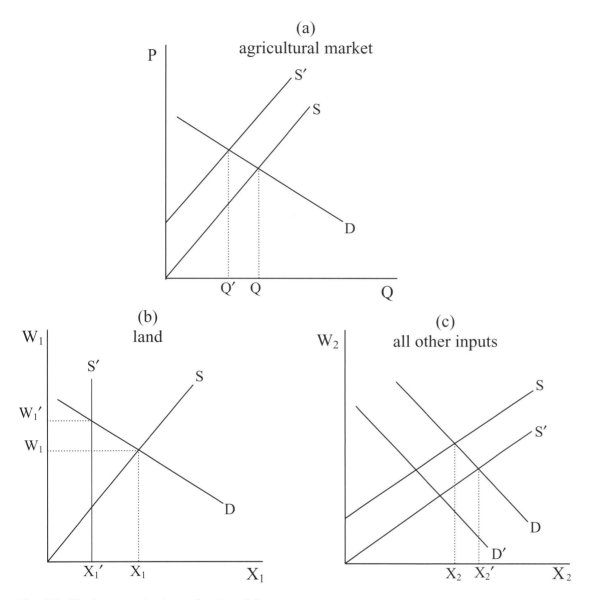

Fig. 7.2. **Simple one output, two input model**.

Algebraically these three markets can be represented by the following system of six equations (Mullen, Alston and Wohlgenannt, 1989; Alston, Norton, and Pardey, 1995, Chapter 4.3.4).

$$Q = g(P), \tag{1}$$
$$P = c(W_1, W_2), \tag{2}$$
$$X_1 = c_1(W_1, W_2)Q, \tag{3}$$
$$X_2 = c_2(W_1, W_2)Q. \tag{4}$$
$$X_1 = h_1(W_1, b_1), \tag{5}$$
$$X_2 = h_2(W_2, b_2). \tag{6}$$

The six endogenous variables in the model are the produced quantity of the agricultural commodity Q and its price P, as well as the two input factor quantities X_1, and X_2 and their prices W_1 and W_2. Equation (1) describes the demand for the agricultural product. Equation (2) states that in the long-run product price equals minimum average total cost. Equations (3) and (4) are derived from the cost function using Shepard's Lemma and are conditional (output constant) factor demand functions. Equations (5) and (6) are input factor supply equations with b_1 and b_2 being exogenous shift variables by which the result of government intervention can be integrated in the model.

As mentioned above, most CSPs either directly restrict the use of input factors, usually land or chemical inputs, or make them more expensive. As illustrated in Figure 7.2b a restriction of land might cause the supply curve to change from S to S' implying displacements in all three markets. Such a set-aside requirement will cause the shadow price of land to increase from W_1 to W_1', causing a leftward demand shift (in the case of gross complements) in the market of all other inputs from D to D' and hence a decline in production from Q to Q' and so on.

Using standard EDM procedures Salhofer and Sinabell (1999) show that the exact effect of a set-aside requirement within the common floor price policy of the CAP is given by:

$$EQ = \frac{k_1(\sigma + \varepsilon_2)}{\sigma + k_1 \varepsilon_2} \beta_1 \leq 0, \tag{7}$$

where EQ is the percentage change in output, β_1 is the percentage change in land caused by a set-aside obligation, k_1 is the cost share of land, ε_2 is the elasticity of supply of all other inputs, and σ is the elasticity of substitution between land and all other inputs. Equation (7) reveals, firstly the direction in which a set-aside scheme pushes agricultural output, and secondly which particular parameters the market effects of CSPs do hinge on. Since under usual assumptions (k_1, ε_2, and $\sigma \geq 0$) all parameters are positive and β_1 is negative (since some land is taken out of production), a set-aside programme will never increase agricultural output.

Further insights are gained by investigating how the change in output induced by a set-aside requirement changes with a change of the forcing parameters and hence by differentiating EQ/β_1 with respect to the parameters:

$$\frac{\partial(EQ/\beta_1)}{\partial \sigma} = \frac{k_1 \varepsilon_2 (k_1 - 1)}{(\sigma + k_1 \varepsilon_2)^2} \leq 0, \tag{7a}$$

$$\frac{\partial(EQ/\beta_1)}{\partial k_1} = \frac{\sigma^2 + \varepsilon_2 \sigma}{(\sigma + k_1 \varepsilon_2)^2} \geq 0, \tag{7b}$$

$$\frac{\partial(EQ/\beta_1)}{\partial \varepsilon_2} = \frac{k_1 \sigma (1 - k_1)}{(\sigma + k_1 \varepsilon_2)^2} \geq 0. \tag{7c}$$

Equation (7a) indicates that an increase in the elasticity of substitution will decrease the ratio EQ/β_1 and hence reduce the effects of a set-aside programme on final output. An increase in k_1 or ε_1 will have the opposite market effect in increasing output.

All Regulation 2078/92 CSPs do have in common that farmers get direct payments as a compensation for the losses they incur when complying with the restrictions on property rights or production possibilities. As described in Section 7.2 most of these payments are decoupled in a sense that they do not depend on the quantity produced. However, they implicitly might have an influence on the quantity produced if some share of these transfers is used to purchase additional quantities of the unrestricted inputs, as discussed in Section 7.3, and hence stimulate production. In such a case the direct payments in fact subsidise the non-restricted inputs. This is depicted in fig. 7.2c by a rightward shift of S to S' in the market of all other inputs implying displacements in the other two markets.

The exact impact of this demand shift on final output again can be assessed by use of the EDM model:

$$EQ = \frac{k_2 \sigma}{\sigma + k_1 \varepsilon_2} \beta_2 \geq 0, \tag{8}$$

with k_2 being the cost share of all other inputs, and β_2 the percentage change in the quantity of all other inputs caused by the compensation payments.

Therefore, final output will never decrease (but is very likely to increase) if some of the direct payments are used to purchase additional units of the non-restricted inputs. The market effect of the reinvested direct payments will be larger the higher k_2 and σ and the smaller k_1 and ε_2:

$$\frac{\partial (EQ/\beta_2)}{\partial \sigma} = \frac{k_1 k_2 \varepsilon_2}{(\sigma + k_1 \varepsilon_2)^2} \geq 0, \tag{8a}$$

$$\frac{\partial (EQ/\beta_2)}{\partial k_2} = \frac{\sigma(k_1 \varepsilon_2 + \sigma)}{(\sigma + k_1 \varepsilon_2)^2} \geq 0, \tag{8b}$$

$$\frac{\partial (EQ/\beta_2)}{\partial k_1} = \frac{-k_2 \varepsilon_2 \sigma}{(\sigma + k_1 \varepsilon_2)^2} \leq 0. \tag{8c}$$

$$\frac{\partial (EQ/\beta_2)}{\partial \varepsilon_2} = \frac{-k_1 k_2 \sigma}{(\sigma + k_1 \varepsilon_2)^2} \leq 0. \tag{8d}$$

In combining the above results the conclusion is that the overall effect of CSPs on agricultural output is ambiguous if it is accepted that some share of the direct payments is used to buy additional inputs. While the restriction of input factors decreases final output, the possible reinvestment of the transfers will increase output. The conditions under which the negative effect of the restriction is larger than the positive effect of the direct payments can be investigated by utilising Eqs (7) and (8):

$$k_1(\sigma + \varepsilon_2) \mid \beta_1 \mid > k_2 \sigma \beta_2. \tag{9}$$

No conclusion with respect to the overall effect of CSPs can be drawn in general, shifting this question into the domain of empirical investigations. This question will be dealt with in the next sections, when the details of the policy for which this method is applied are reported, the parameters that were chosen for the empirical specification of the model are described, and finally the results are presented.

7.6.2. The Austrian 'crop rotation scheme'

The Austrian 'crop rotation scheme' was chosen for this case study because it is among the biggest CSP programmes of the EU in terms of transfers (160 Million ECU in 1996) as well as in terms of acreage (1.2 Million ha).[2] This scheme pertains to the policies restricting input use which is the largest group of CSPs according to the inventory of Deblitz and Plankl (1998), in fact it

---

[2] According to Deblitz and Plankl (1998) there are only four CSP schemes having a broader coverage: Prime à l'herbe (France), Bayerisches Kulturlandschaftsprogramm Teil A (Germany), Elementarförderung (Austria), and Maataloudcn ympäristöohjelma (Finland).

accounted for around 10% of all CSP transfers in the EU in 1995. In addition, the 'crop rotation scheme' is a horizontal measure which is offered across the whole country using a mechanism that does not prevent rents as those discussed in Section 3. It is therefore a good example to show both effects identified above, output reduction due to input restriction, as well as output stimulation in case direct payments are used to buy additional inputs.

In Austria Regulation 2078/92 was implemented with the agri-environmental programme ÖPUL in 1995, the year Austria joined the European Union. This programme offers 25 schemes which cover all elements designated by Regulation 2078/92 with the notable exception of the promotion of 'land management for public access and leisure activities'. The acceptance of this programme (measured as the number of farms enrolled) lies between 170,000 (almost 80% of all agricultural holdings are participating in the scheme 'elementary support') and 0 (the scheme 'reduction of livestock'). The 'crop rotation scheme' is the second most important Austrian scheme in terms of acreage and the most important one in terms of transfers to farmers (BMLF, 1996).

Farmers enrolling this scheme must comply with the following criteria:

- a maximum of 75% of arable land may be used to produce cereals and maize, and
- a winter cover crop (covering at least 15% of arable land) must be planted before 1 November and may not be ploughed under before 1 December.

The effects of the CSP requirements are:

- a decrease in production by restricting the farm owned factor land that can be used for cereals and maize production, and
- an increase of cost for those producers which have to plant winter cover crops because they need more purchased inputs (seed, energy, machinery) apart from increased labour input.

Premiums range from 900 to 1900 ATS/ha (67 to 140 ECU/ha) and are paid according to the share of arable land covered by winter cover crops. The average premiums were 1100 ATS/ha in 1995 (BMLF, 1996) which implies that at least approximately 20% of arable land was covered during the winter season. Some forage crops are defined to be winter cover crops, therefore many livestock producers automatically meet the second criterion.

Part of the premiums are necessary to cover the additional cost of planting the winter-cover crops. The remainder of the premiums increases farm household income which is used for consumption, and/or subsidises the non-restricted inputs. Such additional inputs can be used either for the production of cereals and maize (see Section 7.6.4) or for some other production branch of the farm.

7.6.3. Parameters of the equilibrium displacement model

Values for parameter k_1, k_2, σ and ε_2 (reproduced in Table 7.4) are based on several sources and are derived in more detail in Salhofer and Sinabell (1999). Parameter β_1 ideally would be calculated by dividing land owned by CSP participants actually used for cereal and maize production by land of CSP participants used for cereal and maize production prior to the implementation of the program. Due to data limitation, such detailed information is not available. Instead, β_1 is approximated by the reduction of land used to produce cereals and maize caused by the programme relative to the area before the introduction of the programme which is calculated to be −9 %.

Information obtained from the farm survey provides an indication how farmers spend money they get via public transfers (see Table 7.5). The figures reported here are related to all transfers farmers are getting, not just CSP premiums. These data suggest that only a relative small fraction of transfers is used for consumption and that the major part is used to buy variable inputs and investment goods for agricultural production.

Table 7.4
Parameters used for the empirical EDM model

Parameter	Low	High
k_1	0.20	0.35
k_2	0.65	0.8
ε_2	1	3
σ	0.5	1.5
β_1	−0.09	−0.09
β_2	0.06	0.13
η	∞	∞

Source: Salhofer and Sinabell (1999).

Table 7.5
Average shares of how agricultural transfers are spent—whole sample

EU Member State	Consumption	Variable inputs	Investment	Non-farm investment
Austria	17	46	30	7
Belgium	14	64	19	2
France	5	80	14	2
Germany	5	51	38	5
Greece	9	66	24	2
Italy	22	54	22	2
Sweden	6	44	34	16
United Kingdom	9	65	23	2

Source: STEWPOL farm survey data, Bergström, 1999.
Notes: $N = 1564$. The question was: 'For most farms it is true that they are supplemented by revenues from public funds. Given that your farm is compensated by such extra payments, please indicate to which percentage these transfers would be used for: ...'

If, in general, the argument is accepted that part of the transfers are used to buy additional inputs the implication is that there will be a displacement effect. Unfortunately there is no information on how much of the money farmers receive for enrolling in agri-environmental schemes is actually spent on additional farm inputs and therefore a lower and upper estimate must be assumed in the simulation of the market displacement model.

The range of the second shift parameter β_2 (additional input demand implied by direct payments; see Eq (8)) is derived from a sub-sample of the Austrian data-set to account for the fact that the 'crop rotation scheme' is mainly addressed to crop producers. Farmers participating in the Austrian agri-environmental programme and indicating that crop production is their major source of income report that 40% of direct payments are used to buy variable inputs and 25% are reinvested in durable equipment which is below the Austrian average reported in Table 7.5.

The survey figures apart from 'consumption' appear to be high for several reasons: (i) the spending behaviour may be caused by the fact that some of the respondents may not yet have adjusted their purchasing behaviour to the generally lower price levels on agricultural markets

Table 7.6
Results of the equilibrium displacement model for the Austrian 'crop rotation scheme' in Austria and the EU

	Lower estimate		Mean value		Upper estimate	
	Austria (%)	EU (%)	Austria (%)	EU (%)	Austria (%)	EU (%)
Output effect due to land restriction	−7.1	−0.1	−4.8	0	−2.6	0
Output effect due to direct payments	+1.3	0	+4.5	0	+10.2	0.1
Overall effect on grains output	−5.8	0	−0.3	0	+7.5	0.1

(in particular in Austria and Sweden); (ii) it might be argued that the investments are in fact not output increasing but are made to substitute time that is either used for leisure or for conducting off farm activities (a considerable share of farms is run by part time farmers); (iii) it might be plausible that product revenues otherwise used to buy inputs are used for consumption and direct payments are used to finance input purchases—CSP premiums and product revenues would therefore just offset each other; (iv) as shown in Table 7.2, many participants of CSPs had to make investments as a condition for participation which of course must be covered by the programme. However, participants of the 'crop rotation scheme' do not need to invest in special machinery.

However, from a dynamic perspective the figures of Table 7.5 can be explained if farmers use transfers that are deemed to be income compensations to make new investments in order to increase future income. Hence, such transfers can be distributed over a longer period if farmers assume that contracts are not renewed after five years.

Bearing in mind these facts the survey figures are adjusted downwards considerably and the assumption is made that 20–40% of the direct payments are used to buy additional inputs. The total of direct payments from the 'crop rotation scheme' (BMLF, 1996) multiplied by these percentages, accounting for the additional cost for planting winter-cover crops, and dividing these numbers by the total cost of cereal and maize production, gives a lower and upper bound of β_2 of 0.06 and 0.13, respectively.

7.6.4. Quantitative results of the equilibrium displacement model

Utilising the theoretical results of the EDM model and the described parameter values it is possible to quantify the effects of the set-aside requirement, the effect of the direct payments, and the overall effect on output. Table 7.6 reports upper and lower bounds as well as means of the output effects. The means cannot be calculated directly but have to be derived using Monte Carlo simulation procedures in a way proposed by Krinsky and Robb (1986, 1990) and recently discussed and applied by Zhao *et al.* (1999), Davis and Espanoza (1998) and Salhofer (1999).

To obtain such results, firstly, the ranges of the parameters have to be chosen (Table 7.4), and, secondly, one has to assume a distribution of the parameter values within them. Here, two different distributions are assumed and illustrated for the case of σ. It is assumed that each parameter is either distributed uniformly between the ranges given in Table 7.4 or following a normal distribution around the median of the range ($\sigma = 1$) with a significance level of 95%. Hence, for example one may assume that σ is either uniformly distributed between 0.5 and 1.5 or normally around 1

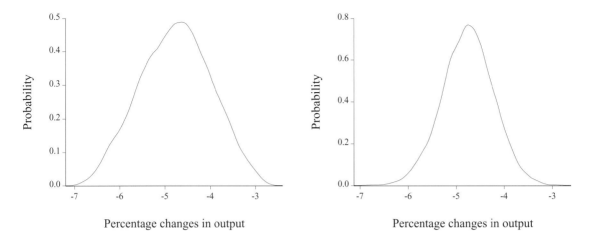

Fig. 7.3. **Kernel density functions of the effect of the set-aside requirement on changes in output assuming a uniform (left) and normal (right) distribution of model parameters.**

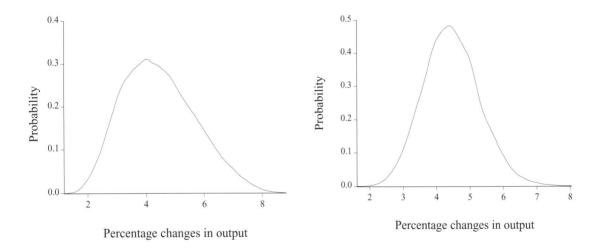

Fig. 7.4. **Kernel density functions of the effect of direct payments on changes in output assuming a uniform (left) and normal (right) distribution of model parameters.**

with 95% within 0.5 and 1.5. By randomly choosing values from these distributions for each parameter, plugging them into Eqs (7) and (8) and repeating this procedure 10,000 times one can derive the mean as well as a distribution of the changes in output.

The means obtained by this procedure are not significantly different between the two alternative distributions of parameters. In Table 7.6 the results of the mean values, the lower, and the upper estimates are summarised. The set-aside requirement reduces the output between 2.6 and 7.1 percent with a mean of 4.8. The additional demand caused by the direct payments increases the output between 1.3 and 10.2 percent with a mean of 4.5. The overall effect is measured to be between −5.8 and +7.5 with a mean of −0.3. The distributions of the changes in output are illustrated in Figures 7.3 to 7.5 using kernel density functions.

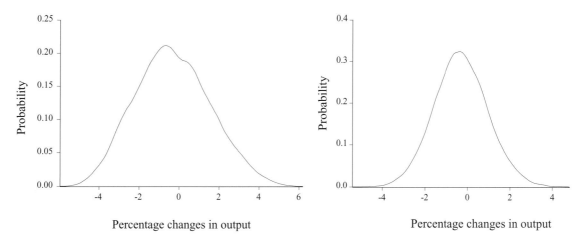

Fig. 7.5. **Kernel density functions of the overall effect on changes in output assuming uniform (left) and normal (right) distribution of model parameters.**

Table 7.6 and Figures 7.3 to 7.5 show that the effect of the restrictions on the farm-owned input land leads to a reduction of output. This effect is smaller than would be expected from the 9% reduction of land which is explained by the substitution of land by other factors. Similar results are to be expected if other inputs than land are reduced. Since Austria produces only a small amount of grains compared to the rest of EU there is practically no effect at the EU level.

If the assumption is maintained that part of the direct payments are used to buy additional inputs output is increasing to almost the same extent. The overall consequence of both effects is that on average no significant output reductions can be observed because of this programme. Even in an extreme scenario where the upper estimates of the 'direct payment effect' and the (absolute) lower estimates of the 'set-aside effect' are combined aggregate effects at EU market levels are not perceptible since Austria produces only a small amount of grains compared to the rest of the EU.

7.7. Summary and conclusions

The goals of Regulation 2078/92 are to reduce farm output and/or to reduce environmentally detrimental side effects of farm production and/or to support farm income. In all EU Member States programmes have been established according to this regulation and over 800 schemes have been introduced. By such schemes the government leases property rights from farmers, usually for a period of five years, or sets up quasi-markets for public goods which are coupled products of agricultural production. The instruments used can almost exclusively be classified to belong to the set of standard instruments of agricultural policy: input restriction, output restriction, and production premiums.

Two major groups of schemes can be differentiated according to their major focus: firstly, schemes which aim at reducing negative environmental impacts of farm production by restricting the use of inputs, and secondly, schemes that aim at the upkeep of agricultural production in marginal areas which frequently use a combination of various policy instruments. This study analyses the output consequences of Regulation 2078/92 schemes of the first type at the market level. To obtain more insight on how agri-environmental programmes actually influence output, the findings of a farm survey were used to obtain information at the programme level. In addition, an equilibrium displacement model was developed to analyse the output effects of the most frequently used schemes, those limiting the use of inputs. This model is applied empirically to the Austrian 'crop rotation scheme' which is among the biggest schemes of the EU both, in terms of transfers, as well as in terms of acreage and which is representative for the group of agri-environmental schemes using input controls.

In the farm survey that was carried out in eight EU Member States almost 2000 farmers were asked about their estimate of the long term output consequences of countryside stewardship policies (CSPs) on their farm. The results from this survey are ambiguous. The survey shows that the majority of farmers participating in agri-environmental programmes does not experience a decline of farm output. In fact, only a few participants actually had to reduce the number of livestock or the share of arable land.

The negative output effect of schemes restricting the use of inputs is confirmed by the equilibrium displacement model. Such schemes will clearly reduce the output at an aggregate level. To what extent this happens depends on the model parameters which must be specified empirically.

A surprising result of the farm survey is that some farmers reported that output actually is going to increase as a consequence of programme participation. Such a result can be explained, if some share of the premiums paid to farmers is used to buy additional, non-restricted inputs. The analysis of the mechanisms that are generally used in CSP schemes shows that at least some participants may obtain rents. Part of these rents actually may be used to buy variable inputs or investment goods.

The 'crop rotation scheme' is a horizontal scheme which is offered across the whole country using a mechanism that does not prevent such rents. It was therefore chosen for a case study to show both effects identified above: output reduction due to input restriction, as well as output stimulation in case direct payments are used to buy additional, non-restricted inputs. The quantitative results obtained from the model analysis are consistent with the findings of the farm survey. The output reducing effect of the restriction on arable land (a requirement of this scheme) may be offset by the output increasing effect of the additional inputs that are financed by the compensatory payments. The final conclusion of the model analysis is that it may well happen that additional inputs, being financed by premiums farmers receive from agri-environmental programmes, may weaken or even cancel out an intended output reduction.

The data obtained from the farm survey and the model results are therefore consistent and suggest that CSPs are not a set of policies from which output reductions can be expected in any case. This obviously reflects the trade-off among the objectives of Regulation 2078/92 which states that apart from output reduction two other goals, improvement of the environment and supporting farm incomes, are equally important. In general, transfers for CSPs—compared to other positions of the EAGGF—are playing only a minor role at the Community level. More than a third of these funds is spent in Austria, Finland, and Ireland which in sum hold only a marginal share in EU agricultural markets. Therefore one conclusion of this study is that market effects of countryside stewardship policies which require participating farmers to restrict the use of inputs are negligible.

The long term output consequences of CSPs may be different. According to Agenda 2000 such policies will become more important in the near future and more funds will be used for them than previously, but such future programmes may be designed differently. It may well be the case that horizontal measures similar to the one described in detail here might no longer be offered. A second aspect is that in future farmers may spend the premiums they get in a different way than they did in 1998, the year when the farm survey was conducted. Currently Regulation 2078/92 measures are limited to a five year contract period but if farmers expect that agri-environmental programmes will be in place for the foreseeable future in particular investment decision may change. A third aspect is that farmers may experience output reduction only in the longer term when the pool of nutrients in their soil declines over time as a consequence of programme participation.

References

Alston, J. M., Norton, G. W. and Pardey, P. G. (1995) *Science under Scarcity: Principles and Practice for Agricultural Research Evaluation and Priority Setting*. Cornell University Press, Ithaca.

Baldock, D. and Lowe, P. (1996) The Development of European Agri-Environmental Policy, In: M. Whitby (Ed.), *The European Environment and CAP Reform*. CAB International, Wallingford, Oxon, pp. 825.

Bergström, P. (1999) *STEWPOL Farm Survey Data File*. Department for Economics, Swedish University of Agricultural Sciences, Uppsala.

BMLF (Bundesministerium für Land- und Forstwirtschaft) (1996) *Grüner Bericht 1995 (Austrian Farm Income Report 1995)*. Vienna.

Davis, G. C. and Espinoza, M. C. (1998) A Unified Approach to Sensitivity Analysis in Equilibrium Displacement Models, *American Journal of Agricultural Economics*, 80, 868–879.

Deblitz, C. (1998) Vergleichende Analyse der Ausgestaltung und Inanspruchnahme der Programme zur Umsetzung der VO, EWG 2078/92 in ausgewaehlten Mitgliedsstaaten der EU. *Arbeitsbericht des Instituts für Betriebswirtschaft*, FAL, Braunschweig.

Deblitz, C. and Plankl, R. (1998) *EU-wide Synopsis of Measures according to Regulation REG, EEC 2078/92 in the EU*. Federal Agricultural Research Centre, FAL, Braunschweig.

Floyd, J. E. (1965) The Effects of Farm Price Supports on the Returns to Land and Labor in Agriculture, *Journal of Political Economy*, 73, 148–158.

Gardner, B. L. (1987) *The Economics of Agricultural Policies*. McGraw-Hill, New York.

Gatto, P. (1999) *STEWPOL CSP Policy Inventory Data Files*. Dipartimento Territorio e Sistemi Agro-forestali Sezione Economia e Politica Agraria e Forestale AGRIPOLIS, Legnaro.

Hofreither, M. F. (1998) *Sustainability in Agriculture—Tensions between Ecology, Economics and Social Sciences*. Paper presented at the Internationale Konferenz der IER und der Universität Hohenheim, 29–30 October 1998, Stuttgart

Howitt, R. E. (1995) Positive Mathematical Programming, *American Journal of Agricultural Economics*, 77, 329–342.

Krinsky, I. and Robb, A. (1986) On Approximating the Statistical Properties of Elasticities, *Review of Economics and Statistics*, 68, 715–719.

Krinsky, I. and Robb, A. (1990) On Approximating the Statistical Properties of Elasticities: A Correction, *Review of Economics and Statistics*, 72, 189–190.

Latacz-Lohmann, U. (1998) Mechanisms for the Provision of Public Goods in the Countryside, In: S. Dabbert, A. Dubgaard, L. Slangen and M. Whitby (Eds.), *The Economics of Landscape and Wildlife conservation*. CAB International, Oxon.

Mullen, J. D., Alston, J. M. and Wohlgenannt, M. K. (1989) The Impact of Farm and Processing Research on the Australian Wool Industry, *Australian Journal of Agricultural Economics*, 33, 32–47.

Muth, R.F. (1964) The Derived Demand Curve for a Productive Factor and the Industry Supply Curve, *Oxford Economic Papers*, 16, 221–234.

OECD (1995) *Adjustment in OECD Agriculture*, Paris.

Salhofer, K. (1999) *Distributive Leakages from Agricultural Support. Some Empirical Evidence from Austria*. Paper presented at the 1999 Annual Meetings of the Southern Agricultural Economics Association in Memphis, Tennessee.

Salhofer, K. and Sinabell, F. (1999) Utilising Equilibrium-Displacement Models to Evaluate the Market Effects of Countryside Stewardship Policies: Method and Application, *Die Bodenkultur*, 50, 141–151.

Scheele. M. (1996) The Agri-environmental Measures in the Context of the CAP-Reform. In: M. Whitby (Ed.), *The European Environment and CAP Reform*. CAB International, Oxon, pp. 3–7.

Weaver, R. D. (1998) Private Provision of Public Environmental Goods: Policy Mechanisms for Agriculture. In: S. Dabbert, A. Dubgaard, L. Slangen and M. Whitby (Eds), *The Economics of Landscape and Wildlife conservation*. CAB International, Oxon.

Vogel, S. (1998) *Umweltbewußtsein und Landwirtschaft. Theoretische Überlegungen und Empirische Befunde* (*Environmental Attitudes and Agriculture. Theoretical Considerations and Empirical Findings*). Habilitationsschrift, Institut für Wirtschaft, Politik und Recht, Universität für Bodenkultur Wien.

Zhao, X., Griffiths, W. E., Griffith, G. R. and Mullen, J. D. (1999) Probability Distributions for Economic Surplus Changes: The Case of Technical Change in the Australian Wool Industry, *Australian Journal of Agricultural and Resource Economics*, forthcoming.

G. Van Huylenbroeck and M. Whitby
Countryside Stewardship: Farmers, Policies and Markets
© 1999 Elsevier Science Ltd. All rights reserved

Chapter 8

Analysis of trade distortions

Dimitri Damianos and Yanni Barlas

Abstract—Environmental issues related to agricultural trade are gaining growing importance in recent years, particularly in terms of the effects of trade liberalisation on the environment and the trade-distorting implications of agri-environmental measures. Within this context, the Uruguay Round Agreement allows for the implementation of agri-environmental support under certain conditions. The objective of this chapter is to assess the trade-distorting effects of agri-environmental policies implemented in eight EU Member States. Two approaches are followed for the assessment of trade-distorting effects of nine agri-environmental policy groups, namely the Producer Support Estimate (PSE) concept and a complementary qualitative approach which assesses the cost effectiveness of them in providing environmental benefits on the basis of the transparency, targeting, tailoring, evaluation and monitoring criteria. The PSE analysis shows that agri-environmental measures in the EU can be regarded as minimally or even non-trade distorting, as their share in total PSE remains far below the critical 5% threshold. Similarly, the qualitative analysis shows that trade distortion is rather low, though more attention should be attached on the criteria of targeting and tailoring. Finally, wildlife and biodiversity policies show a very satisfactory level of adherence to all criteria, while popular policies such as reduction of negative impacts from agriculture and recreation and access to agricultural land do not perform satisfactorily.

8.1. Introduction

Trade and trade-policy implications from linking agriculture and the environment are currently receiving growing attention, particularly in terms of issues such as the effects of trade liberalisation on the environment and the possible trade-distorting implications of agri-environmental support measures. Furthermore, the increased emphasis in the formulation and implementation of agri-environmental measures and programmes at the EU-level, constitutes an effort on behalf of a major actor in the international economy to integrate environmental considerations in agricultural policy design and implementation.

Recent developments in the field of international trade, such as the establishment of the World Trade Organisation (WTO) and the adoption of the relevant multilateral agreements in the Uruguay Round negotiations, have set a new context for the formulation and implementation of agricultural policies including the possibility of incorporating agri-environmental objectives and measures into agricultural policies. Also, the Uruguay Round agreement contains provisions (such as the specification of several environmentally-friendly measures as 'decoupled') that relate to the environment, allowing for the implementation of agri-environmental support under certain conditions, which will probably become more restrictive in the future.

Within this context, this chapter attempts to identify and assess the possible trade-distorting effects of agri-environmental policy measures implemented in the STEWPOL countries. Two approaches are followed for this purpose, namely the calculation of the effect of agri-environmental policies on the Producer Support Estimate (PSE), which is regarded as an index measuring trade distortion (OECD, 1987), and the use of a qualitative approach which analyses how far selected CSP measures meet the conditions of non-distortion (OECD, 1998).

The chapter is organised as follows: Section 8.2 presents the theoretical context regarding the effects of agri-environmental policies on trade, while Section 8.3 reviews the agri-environmental aspects of the Uruguay Round Agreement. Section 8.4 presents the methodology used for the analysis of the trade-distorting effects of agri-environmental policies (qualitative approach), while the relevant results are reported in Section 8.5. Finally, conclusions are presented in Section 8.6.

8.2. The relationship between agri-environmental policies and trade: theoretical considerations

8.2.1. General framework

Agricultural activity is, by its nature, closely related to the environment, using it as a factor of production and a source of inputs, and making a deliberate technological use of soil and biological processes in producing its output. In this context, agriculture can have positive or negative effects on the scenic quality of landscapes, on the diversity of ecosystems and thus produce either a 'public good' (such as environmental quality) or 'negative externalities' (such as the pollution of groundwater or the destruction of a desirable landscape feature).

Farming practices are the main vehicle via which agricultural activity exerts influence on the environment. Changes in farming practices, changes in the output mix and/or in the mix and intensity of inputs used, do have implications for the state and quality of the environment. From their part, farming practices are greatly influenced by agricultural policies implemented. Therefore, agricultural policies may enhance or diminish any positive or negative environmental effects of agricultural activity.

This brief sketch of the linkages between agriculture and the environment underlines the fact that the relevant problems are the combined result of 'market failure' on the one hand, and of 'policy or intervention failure' on the other. In principle, such impacts would be kept at socially acceptable levels if market forces could effectively reflect social costs and benefits in restoring environmental quality that was lost due to agricultural activity.

International trade may, also, influence the environment, but mainly indirectly. Trade affects the scale and geographical pattern of production, increases consumption possibilities and influences the direction and rate of technological innovation and diffusion. It also permits the location of production to be separated from the point of consumption and therefore affects the environment if it causes agricultural production to shift from places where it is less sustainable to places where it is more sustainable or vice versa. In addition, trade stimulates economic growth and development and thus increases welfare. Therefore, more financial resources could become available to tackle environmental problems but also with an increased tendency to intensify production, causing more negative effects.

Since environmental problems are basically the result of 'market and intervention failures', trade liberalisation and agricultural policy reform do not seem to be able to resolve all agri-environmental problems. Trade liberalisation may contribute to a reduction of certain environmental problems related to 'intervention failure', however, problems related to 'market failure' will still persist. Therefore, an active and effective environmental policy (i.e. government intervention) to correct 'market failure' seems inevitable either in an incentive approach form or in a regulatory one (OECD, 1998).

8.2.2. Trade effects of agri-environmental policies: theoretical considerations

Agricultural activity is characterised by wastes and/or amenities, produced as by-products, which in turn constitute environmental externalities, negative or positive respectively, in the sense that the farmers are not charged with the cost of pollution abatement, neither are they compensated for the amenities they provide society with. Producers maintain activities, which pollute while they fail to adopt technologies which are beneficial to the environment, because such external costs are not incorporated into their private production costs. Such market failure or market imperfection calls for government intervention and the implementation of corrective measures.

Agri-environmental policies thus affect investment, technological change, production, consumption and trade (Pearce and Turner, 1990). In selecting environmental policy instruments in agriculture, the state decides on how much pollution should be reduced as well as on what type of policy is most effective in achieving that goal.

A theoretical model for defining the optimal pollution level has been adopted by Krissof *et al.* (1996), utilising the notions of marginal external costs which measures the additional costs of pollution to society from changes in production levels and net private benefits which reflects the benefit minus private cost accruing to the farmer from different levels of production, in order to describe the value society places on the damage from pollution. The optimum level of economic activity and pollution for the society occurs at the level of production where the marginal benefit to producers equals the marginal costs of pollution to society. According to the model, if producers were obliged to pay a tax on production, the optimal production and pollution levels would be maintained (Polluter Pays Principle—PPP).

By means of the same theoretical model, Tobey and Smets (1996) have tried to assess whether the EU agri-environmental Regulation 2078/92 is consistent with the Polluter Pays Principle (PPP), arguing that related payments impose environmental performance standards as expressed and enforced by the rules of good farming practices, that do not meet the socially optimum level of environmental quality, but instead, an inferior optimum level (weak version of the PPP). Thus, financial incentives are provided to farmers, with the ultimate aim to improve the environmental quality without introducing a large cost burden on them. However, farmers are requested to pay the cost of pollution abatement up to the point that denotes good farming practices only by confirming to benchmark regulations, but may receive pollution-abatement compensation beyond that point.

In order to determine whether a subsidy for pollution abatement beyond the benchmark is a valid one or not, practical information on what constitutes a good agricultural practice is required. Such information needs to be adopted according to national, regional and local characteristics and conditions. Payment schemes however are uniformly implemented as it is practically impossible to determine for each particular case, what is a change in agricultural practice and what is already an existing practice (Tobey and Smets, 1996).

In theory, agri-environmental measures, such as the imposition of an environmental tax or conversely the granting of a compensation payment, will reduce domestic production, pollution and exports. In the case of a small country exporter incapable of affecting world prices, the above effects stand, while domestic consumption levels are likely to remain unchanged as consumers are confronted with an unaffected world price. However, if an agri-environmental measure (tax) is applied in a country, which is a large trader of an agricultural commodity, prices also rise for both domestic and foreign consumers, due to supply reductions. Naturally, stringency in policy implementation, along with the nature of measures imposed, determine the direction and magnitude of such effects.

It is worth noting here that EU agri-environmental policy allows for a high degree of flexibility when it comes to applying specific measures in the different Member States. Therefore, different effects on the direction and composition of trade can appear. The extent of the application of an environmental measure is another crucial factor. If agri-environmental regulations are confined within limited boundaries and sites, address location-specific environmental problems or are targeted to a limited number of producers, effects on trade are negligible, as competitiveness at the national level remains unaffected. The analysis presented here does not take into account technological change likely to come about in the long run, that can eventually bring about the desired environmental quality without affecting agricultural production and trade. Alternatively, agri-environmental measures, may affect production cost and hence the level of production and trade in the short run.

8.2.3. Evidence

Evidence, so far, suggests that the effect of applied agri-environmental measures on trade are almost negligible. The voluntary subsidy approach which characterises agri-environmental policy, e.g. in the EU has little to no effect on production cost and trade competitiveness with the possible exceptions of land set-aside and certain regulations on the use of chemicals (OECD, 1997, see also Chapters 6 and 7).

This generalisation is based on the fact that pollution abatement costs and consequently compensation and subsidies comprise a small share of the industry's cost. The issue of trade effects in agriculture, as a consequence of agri-environmental programme implementation, is a very recent one. Although OECD favours the application of the PPP, the EU uses subsidies, on the grounds that European farmers have historically and unconditionally retained rights in the use of natural resources and therefore, they require compensation in return for the diminution of such rights. However, if such subsidies are restricted to the minimum level necessary to include changes in farm production and are tailored to encourage innovation of achieving environmental improvement, they are not trade distortive (OECD, 1997).

Also, Tobey (1991) has suggested that the magnitude of trade competitiveness losses due to the implementation of agri-environmental measures is quite modest, as most major participants in the agricultural world trade have introduced similar agri-environmental programmes, the trade effects of which are counterbalanced and thus eliminated. On the other hand, countries which do not apply agri-environmental programmes do not hold large market shares in most agricultural commodities. Lastly, the presence of larger forces such as labour costs, exchange rate fluctuations, distribution networks and mostly price and income support measures are the major determinants of any change in trade competitiveness.

In order to evaluate agri-environmental policies in terms of PPP and assess potential distortive effect on competitiveness and trade, one first has to identify and qualify all the various types of environmental subsidies. Literature suggests that the magnitude of such subsidies is not large (Tobey and Smets, 1996). Although the exact magnitude of environmental subsidies that can be held responsible for trade distortions is not specified in the final act of the Uruguay Round of Multilateral GATT Negotiations, a subsidy which exceeds 5% of the value of production is supposed to bring about a 'serious prejudice' (Article 6—see below).

Agri-environmental subsidies' impacts on trade could be aggravated in the future, because as production subsidies decline, environmental subsidies in agriculture may follow the opposite trend (e.g. Agenda 2000). The aggregate cost attached to the implementation of agri-environmental measures could therefore have a clear impact on a country's competitiveness when it is compared with countries which impose less regulation and spend less on compensation subsidies.

Current trends in the development and application of agri-environmental programmes indicate that regulation for pollution abatement, whether it implies compensation and subsidies or not, could become a major factor in distorting agricultural trade. It seems that the degree of such distortion will highly depend on the design, nature and extent of future agri-environmental programmes. Another core issue is whether environmental programmes conflict with agricultural trade liberalisation. To answer this, total present compliance costs and the nature of currently applied programmes have to be investigated.

8.3. The agri-environmental aspects of the Uruguay Round Agreement

8.3.1. Main elements and implications

The Agreement on Agriculture (AOA), concluded in the framework of the Uruguay Round Multilateral Trade Negotiations, has been a major turning point in the evolution of agricultural policies and world agricultural trade. The Agreement contains provisions that put discipline on both trade policies as well as on domestic support. More precisely, rules and commitments undertaken cover three broad areas of agricultural and trade policies:

- **Market access** (articles 4, 5 of the AOA) i.e. the rules and concessions, contained in the country schedules, governing the protection against import competition.
- **Domestic support** (articles 6, 7, Annexes 2, 3 of the AOA) i.e. the rules and concessions, contained in the country schedules, relating to the use and the level of 'non-border' measures implemented to support agricultural production.
- **Export competition** (articles 3.3, 8, 9, 10, 11 of the AOA) i.e. the rules and concessions, contained in the country schedules, relating to the subsidisation of agricultural exports.

Following the conclusion of the Uruguay Round Multilateral Trade Negotiations, several attempts on the assessment of the impact of the AOA have indicated that the degree of liberalisation of agricultural trade achieved is rather limited, but the importance of the AOA is great, since it brings agriculture into GATT regulation, it makes border protection measures more transparent, it turns agricultural policies towards direct income support measures, it reduces the possibility of export subsidies and finally, it facilitates future negotiations. On the other hand, analyses have shown that tariffication is not expected to exert significant influences on trade flows and on agricultural product prices in the coming years, mostly due to the fact that most countries have already set the tariff equivalents, for most products, at levels higher than the actual ones in the base period. Finally, significant social welfare benefits have been forecast for the developed world (USA, EU and Japan), while developing countries (especially net food importers) are expected to loose, due to subsidy restrictions in both the USA and Western Europe (Goldin and van de Mensbrugghe, 1995; Harrison et al., 1995).

In sum, the degree of liberalisation obtained by the agreement is expected to be rather moderate. In that sense the agreement could be characterised as a 'partial liberalisation' agreement. It is not expected to have any significant influence on global trade. However, there will be shifts in trade flows as well as in production patterns. The influence of the agreement on the level and variability of world prices is expected to be moderate. Nevertheless, the agreement is significant as it contains innovative elements with important and permanent consequences with respect to the choice of the policy mix.

Turning back to the issue of relations between environmental policies and trade, it is worth referring in some detail to the restrictions on domestic support as imposed in the AOA, as those are mostly related to the implementation of agri-environmental policies.

More analytically, two kinds of restrictions on domestic support were agreed:

(a) **Indirect restrictions**, which are related: first, to quantitative or other commitments on the level of market access (tariffication, reduction on the tariff and tariff equivalent level) on the one hand, and on the competition level of exports (reductions on subsidised exports) on the other; second, to the obligation that, in the framework of the 'peace clause', the annual product-specific support that is given via 'non-decoupled' and 'quasi-decoupled' measures does not exceed the level of support specified for the 1992 marketing year.

(b) **Direct restrictions,** which relate to qualitative and quantitative commitments on the domestic support measures. In this case it is important to distinguish between those that constitute the **'Green Box'** (decoupled) and the **'Blue Box'** (quasi-decoupled), for which no reduction commitments are undertaken, and those that constitute the **'Red Box'** (non-decoupled), that are subject to reduction.

'Decoupled measures' (green box) include all measures and policies that have zero or minimum effect on production and trade, and are explicitly exempted from reduction. In addition, according to Art. 13 ('peace clause'), domestic support measures that conform fully to the provisions of Annex 2, cannot cause the application of countervailing duties and shall be exempted from actions based on the Subsidies Agreement and on non-violation, nullification/impairment of benefits of tariff concessions to another member.

Annex 2 of the AOA determines the 'decoupled' measures and policies, as well as the general and specific decoupling criteria. These policies have been exempted from commitments to reduce domestic support. In terms of environment-related measures, it includes 'Resource Retirement Programmes' (retirement of land for a minimum of 3 years), 'Direct Payments under Environmental Programmes' (fulfilment of specific conditions, including those related to production methods or inputs; payment limited to extra costs or loss of income involved), 'Direct Payments to Producers' and 'Regional Assistance Programmes'.

However, perhaps even more important (in terms of relationship with agri-environmental policies) are the explicit provisions of article 6 on **'Quasi decoupled measures'** (**blue box**) that are also excluded from any reduction commitment. Measures related to production reduction programmes (article 6, para. 5) are calculated in connection with the amount of support for each product, according to the 'peace clause' provisions. Here, article 6 of the AOA provides exemptions from the Aggregate Measure of Support (AMS) and the Equivalent Measure of Support (EMS) calculations for:

- Product specific support which amounts to less than 5 per cent of the value of production of the product concerned, and non-product specific support which amounts to less than 5 per cent of the total value of a country's agricultural production (de minimis provision);
- Direct payments under production-limiting programmes, if such payments are based on fixed area and yields for crops and fixed number heard for livestock products or payments made up to 85 per cent or less of the base level of production (blue box).

Finally, the concept of 'non-decoupled' payments (red box) includes all policy measures which affect producer decisions on what, how much and how they will produce. All measures of this category are summarised by the AMS indicator.

The Uruguay Round Trade Negotiations proved to be lengthy and complicated. Although solutions have been finally found for many issues, others remain to be addressed and future negotiations for continuing the agricultural policy reform and liberalisation of agricultural trade will be initiated this year. Except from agriculture, it is expected that the new Round will focus on many important issues, among which is that of the environment.

Among the purely agricultural issues, pressure is expected for further reductions on tariffs and tariff equivalents, as well as for the enlargement of minimum access opportunities in order to achieve a greater market access. The 'Special Safeguard Clause' is also expected to gain attention, and further reduction of export subsidies will be discussed, aiming at a deeper and more stringent discipline. Concerning domestic support, it is not only its overall reduction that will be on the centre of future discussions. Additional areas of concern include product by product fixing of commitments (and the subsequent reduction in existing flexibility), re-examination of decoupled payments and of their decoupling criteria, and the existence of income support measures, including deficiency payments that relate to supply management measures.

Further to the above, the relationship 'agriculture-environment-trade' is expected to become a main issue. The emphasis which, to date, is put on the protection of the environment in relation to agriculture and trade, is expected to create new areas for agricultural policies as well as new grounds for the debate between: those who believe that free trade and environmental policies can work in tandem to achieve social benefit, economic growth and environmental quality; and those who insist that there is no simple or automatic link between trade liberalisation and environmental protection and that market mechanisms could only lead to both an economically and ecologically optimal allocation of production resources if full internalisation of environmental costs were achieved. Future negotiations will have to combine environmental protection and international competitiveness, i.e. to distinguish between different agri-environmental policies and their trade effects, as well as between different trade measures that secure environmental policies. The aim will be to examine how full implementation of Uruguay Round commitments has the potential to yield benefits for both the multilateral trading system and the environment.

Along these lines, environment-related provisions of the AOA (particularly these included in the Green Box), are expected to come under scrutiny. In more detail:

(a) **Resource retirement programmes** are implemented in different ways (rotational/non-rotational) and the land set-aside faces alternative uses, with (as a result) different environmental impacts. This might lead to a re-examination of the current condition of not having to specify alternative uses.

(b) **Direct payments under environmental programmes** depend for their success on the degree of targeting, spatial distribution and farmer participation level. As farmers seem to participate when benefits exceed costs, these payments seem to face compatibility difficulties with AOA provisions, since the payment cannot exceed the additional cost or revenue foregone relating to compliance.

(c) **Direct payments to producers** are a powerful instrument for securing environmental benefits as public good that arise from farming. However, being delivered as a by-product of food production (multiplicity of objectives), and taking account of the trend for reducing agricultural support, there is a danger that these provisions can no longer be secured. These payments have to be considered as reflecting the value of the 'public good', giving rise to evaluation difficulties and a danger of 'policy failure'. If this happens, then specific programmes might not be characterised as decoupled, since they relate to the type of production.

(d) **Regional assistance programmes** are available to all producers in disadvantaged regions, contributing to the continuation of agricultural activity and the maintenance of rural population. However, as farming in these areas is not economically-sustainable, and market liberalisation is expected to have adverse impacts for the rural environment, important questions on the territorial coverage of this measure are expected to arise.

8.4. Methodology for measuring trade distortion

8.4.1. Alternative approaches

The proposition that environmental policies may affect trade and trade flows, have initiated several research efforts to develop methodologies for assessing likely trade distorting effects. However, in the case of measuring the trade distorting effects of agri-environmental policies, important difficulties arise, since any attempt to propose a methodology for their measurement requires:

(a) a clear definition of what the non-distorting situation would be. Externalities, by definition, are the source of distortions in the allocation of resources both at the national as well as at the international level. Hence, the reference level should correspond to the free world market when all externalities have been properly and accurately internalised.
(b) the establishment of a clear distinction between the harmful and beneficial environmental effects of agriculture which is certainly not an easy task. At present, any judgement has to rely on the nationally determined benchmarks which may be considered as a second best solution especially if these are based upon sound scientific evidence and are closely related to the environmental conditions in each country and/or region.
(c) an evaluation of the various agri-environmental instruments used, in relation to their optimal level. Setting environmental instruments at a level deviating from the optimal one might lead to trade distortions by affecting resource allocation.

The above discussion highlights some of the difficulties in assessing the likely trade distorting effects of agri-environmental measures and indicates that additional work is needed both at the theoretical as well as at the empirical level before any firm conclusion can be drawn.

However, taking into account concerns with regard to the likely trade distorting effects of agri-environmental policies and measures, the use of proxy indicators along with a set of criteria or guidelines may be proved useful for obtaining a rough assessment. A number of single summary indexes have been used or proposed for the domestic and international surveillance of agricultural policy measures. The most popular is the Producer Support Estimate (PSE), which is defined as an indicator of the annual total monetary transfer to agricultural producers from domestic consumers and taxpayers as a result of agricultural policies (Cahill and Legg, 1990, OECD, 1995). However, the PSE suffers from weaknesses, as it does not make any distinction between the different measures by assigning weights according to the degree of trade distortion, while it also does not measure the effects that agricultural policy measures might have on production, consumption, trade and other variables since it simply adds components which may have very different effects on production, trade etc. (Hertel 1989).

In this work, two complementary (one quantitative and one qualitative) methodologies are utilised for assessing the possible trade distorting effects of agri-environmental policy measures implemented in the eight STEWPOL states. One is the calculation of the effects of these policies on the (quite popular, despite its weaknesses) PSE indicator (quantitative method). The qualitative method used derives from OECD (1998) and is based on the identification of the characteristics of agricultural policies which can achieve environmental objectives with least-distorting effects for agricultural markets. The main elements of these two approaches are described beneath.

8.4.2. The Producer Support Estimate (PSE)

The PSE indicators developed (initially as Producer Subsidy Equivalent) by Josling (FAO, 1973) and elaborated by the OECD (1987) are a key element in the annual monitoring of policy developments in the OECD member countries. Although the PSEs are not indicators as such of the degree of protection attributed to agricultural products they are an essential input in the negotiations. Rather they offer a broader measure of assistance provided through the components of a given set of agricultural policies.

The PSE is defined as an indicator of the annual monetary value of gross transfers from consumers and taxpayers to support agricultural producers, measured at the farm gate level. The emphasis in the PSE is placed on the notion of transfers to producers and not on trade distortions arising from agricultural policies (including agri-environmental ones). It does not distinguish between different measures, reflecting thus the underlying assumption that in the long run, **all** measures may influence resource allocation and thus lead to trade distortion.

Another shortcoming of the PSE is that it does not cover certain measures (sub-national expenditure, tax concessions, credit subsidies) and expenditures financed outside the Ministries of Agriculture, as well as other policies that are not considered as transfers to producers.

The PSE indicators are calculated on a product-specific basis. As a monetary measure it permits the monitoring of support over time. It is thus used as a monitoring device to analyse the evolution of agricultural support and can be decomposed to facilitate the evaluation of year to year changes attributed to various support measures. Measures and policies included in PSE calculations include: (a) market price support; (b) payments based on output; (c) payments based on area planted or animal number; (d) payments based on historical entitlements; (e) payments based on input use; (f) payments based on input constraints; (g) payments based on overall farming income; and (h) miscellaneous payments.

In the framework of our research interest, PSE indicators are expressed as a percentage of the value of production, adjusted to include direct payment and exclude producers levies:

$$\text{PSE \%} = \text{PSE}_{Ti} / \{Q_i * P_{di} + D_i - L_i\}$$

where, T = a sub-sample of total payments, for example agri-environmental payments
i = product
P_d = observed domestic price
Q = actual level of production
D = direct payments
L = levies on producers

Regarding agri-environmental measures and from a practical point of view, a benchmark beyond which these transfers may be considered as trade distorting is required. As theory does not offer an indication of this, the de-minimis provision of the AOA was used. According to it, there is minimal, or no trade distortion if:

(a) product specific domestic support does not exceed 5 per cent of the total value of production of the product concerned; and
(b) non product specific domestic support does not exceed 5 per cent of the value of total agricultural production.

Thus, in this work, this level is used as a benchmark to assess the danger for trade distortion, that arises from various agri-environmental measures.

8.4.3. The qualitative approach

In general, it could be argued that there are at least three cases in which 'green subsidies'—including those provided by Regulation 2078/92—may be considered as trade distorting:

- The 'green subsidies' do not finance an environmental outcome;
- The 'green subsidies' do finance an environmental outcome, but producers are being overcompensated, and
- There is no overcompensation, yet 'green subsidies' are not being provided in a cost-effective way.

For any of the above situations, it could be argued that the relevant measures cannot be included under the 'green box' provisions of the UR-AOA. Within this framework, a set of criteria, developed by the OECD (1998), could be utilised to evaluate, in qualitative terms, the likely degree of trade distortions accruing from the agri-environmental measures implemented by the EU. This approach could also be used to evaluate the current policy in relation to its trade distorting effects, as well as to evaluate the relevant policies of other EU trade partners. This qualitative approach is measure- and not goal (objective)-oriented. It is expected thus, that it will provide an assessment of whether various EU agri-environmental policies are rightly included in the 'green box' of the Uruguay Round Agreement on Agriculture as non distorting policies.

According to the OECD criteria, in order to ensure that payments are cost effective in providing environmental benefits, and do not distort agricultural markets, they need to be:

(a) **transparent** in their objectives and operation;
(b) **targeted**, in order to ensure the provision of benefits, which would not be otherwise provided above the recognised reference level;
(c) **tailored** to the particular environmental situation, limited to cover the costs of ensuring the provision of the amount of the benefits desired and accompanied by adequate advice and information;
(d) **evaluated** as to their environmental effects, the results of which would feed back into the possible adaptation of programmes to ensure that environmental needs are being met through alternative lower cost solutions, and
(e) **monitored**, in order to ensure compliance and cost effective implementation.

The first step which concerned the setting of the above criteria was followed by the assignment of weights to each one of them. It was decided that equal weight should be assigned to each one of the criteria, though several questions were raised. The problem seems to arise especially with respect to the evaluation and monitoring criteria which at first instance may be perceived as less important for assessing the trade distorting effects of the measures applied. For the sake of global assessment, it was assumed that the criteria are interrelated and complementary. They should not be seen as single unrelated devices, of different weights or significance. For instance the very basic aim of monitoring is to ensure that producers fully comply to the specific commitments they have undertaken. In the absence of monitoring and/or evaluation there is a real danger that the producers will probably perceive payments, incorrectly, as another way of agricultural production support. Consequently, such a behaviour would have repercussions on both environmental effectiveness and cost efficiency of the measures as well as on the allocation of resources and trade. Thus, despite any bias that may be introduced in the overall assessment, it was preferred to maintain the equal weight assumption.

A further question was raised with respect to whether a scale that could distinguish between various degrees of trade distortion instead of a plus/minus or a yes/no indication should be allowed for. Although the issue was already of concern, this possibility was originally ruled out, with an aim to come back at the end and construct a scale that would allow for distinguishing the likely degree of trade distortion associated with the specific policy measures and the way they are implemented. It was therefore suggested that all answers are classified as 'strongly agree' (SA) 'agree' (A) 'undecided' (U) 'disagree' (D) 'strongly disagree' (SD). Such a classification is useful especially in certain (ideal) cases when ranking is possible on the basis of objective observations. However, the distinction between 'strongly agree' and 'agree' classification may not always be straightforward. When this is the case, the distinction was ignored by indicating both classifications on either side. Also, the sources of any ambiguity in the discussion section of the completed tables were stated.

For each one of the five criteria, a number of sub-criteria (hypotheses) were identified, while answers were classified for each sub-criterion, in the range of the five possible options spelled out above. These hypotheses were as follows:

A. Transparency criterion
1. Payments directly linked to the environmental outcome expected;
2. Payments directly associated with farming practices and/or factors of production;
3. Total payment calculated for each product on the basis of total area cultivated or volume of production or inputs used;
4. All farmers in the defined area are eligible if they can contribute to the provision of the environmental outcome;
5. Clear and detailed definition of instrument.

B. Targeting criterion
1. The specific measure addresses one main explicit environmental objective;
2. Existence of a specific reference level that ensures attainment of the desired environmental outcome;
3. A specific environmental outcome over and above the reference level is intended/expected;
4. The specific environmental benefit desired would not be attained without the implementation of the particular agri-environmental measure.

C. Tailoring criterion
1. Payments adjusted according to prevailing actual environmental, agronomic and other conditions; uniform application justified on sound economic grounds;
2. No evidence of overcompensation for farmers.

D. Evaluation criterion
1. Evaluation procedures have actually been put into effect, concerning: a) the environmental output, b) the economic efficiency of the measure;
2. The measure leads to a positive, quantifiable environmental output, as intended;
3. The intended environmental output is efficiently obtained.

E. The Monitoring criterion
1. Monitoring procedures on compliance of the beneficiaries with the terms of the contract are implemented;
2. According to monitoring procedures implemented, compliance of beneficiaries is assured;

3. Measure is implemented in a cost-effective way, that keeps implementation-cost to the minimum.

However, it should be noted that the relative complexity of the questions and thus the rigour with which they were answered gives rise to the possibility of bias caused by judgement values in the data used.

8.5. Results of the quantitative and qualitative analysis

8.5.1. Quantitative analysis

Agri-environmental measures of Regulation 2078/92 do not belong to a single PSE component, but are assorted to certain categories and sub-categories according to their nature. In more detail, *Organic Agriculture* and *Reduction of Fertilisers* are under (c) Payments based on Area or Animals, since it concerns payment to producers, conditional on production of a specific commodity(ies). *Conversion of Arable Land to Grassland and Forage, Training Demonstration and Soil Improvement* provide payments to producers using specific inputs and are classified under (e) Payments based on input use. *Extensification of Livestock* is based on input use constraints, and are thus classified under (f) Payments based on input constraints (Table 8.1).

Extensification of Livestock had the largest participation in the EU-15 PSE in 1998, while agri-environmental programmes on *Soil Improvement* had the lowest. Total spending on Regulation 2078/92 has shown an impressive increase during the period between the years 1995 and 1998.

As measured by the percentage PSE, support to producers in the EU fell from 46 per cent in period between the years 1986 and 1988, to 39 per cent in the period 1996–98. On the other hand, the share of agri-environmental measures in total PSE increased significantly from 0.2 per cent in 1993, to 2.4 per cent in 1998. In general, although the share of market price support fell, there was a substantial increase in the share of payments based on area planted and animal numbers. Payments based on input constraints (including environmental constraints) increased six fold, but accounted for only 4 per cent of total support to producers at the end of that period. On the other hand, the share of support based on input use remains stable.

Main contributors to this impressive increase of the share of Regulation 2078/92 in the total PSE are measures such as Reduction of Fertilisers, Extensification of Livestock, Public Land Leisure and Upkeep of Abandoned Land, and Conversion of Arable Land to Grassland and Forage. On the other hand, the total PSE regarding Training and Demonstration increased less rapidly, while that on Organic Farming and Soil Improvement, declined.

Total PSE in 1998 increased by 20 per cent (compared to 1997), and the percentage PSE by 7 points to 45 per cent, due to an increase in market price support. This development is attributed to a sharp fall in world prices, the decline in per unit budgetary payments and the depreciation of the ECU against the US dollar. On the other hand, the share of Regulation 2078/92 fell from 2.7 to 2.4 per cent in the same period.

In summary, taking account of the AOA provisions, agri-environmental measures in the EU can be considered as being minimally or even non trade distorting, as their share in total PSE remains, as yet, far below the critical threshold of 5 per cent.

8.5.2. Qualitative analysis

In the context of the methodological framework described in section 8.4.2, 113 policies out of 148 contained in the inventory of this work were evaluated for trade distortion using the OECD quali-

Table 8.1
Total Support Estimate—Aggregate P.S.E. of all Commodities: 1993–1998 (in billion ECU)

Policy Measures	EU—12		EU—15			
	1993	1994	1995	1996	1997	1998[1]
I. Total Value Of Production (at the farm gate)	197,563	205,860	207,397	208,945	210,506	212,077
II. Producer Support Estimate (PSE)	101,878	100,209	98,655	84,481	94,771	114,687
A. Market Price Support	71,756	66,508	59,168	43,449	51,688	71,852
B. Payments Based on Output	4,337	2,633	2,074	3,496	3,632	3,776
1. Based on Unlimited Output	609	408	121	122	95	121
2. Based on Limited Output	3,729	2,225	1,953	3,373	3,537	3,655
C. Payments based on Area/Animals	15,391	19,742	23,538	23,392	25,268	25,198
1. Based on Unlimited Area/Animals	1.083	467	539	844	724	746
of which:						
Env. (2078/92), Organic farming	–	–	–	*207*	*112*	*121*
Env. (2078/92), Land management to reduce the use of fertilisers	–	*1*	*30*	*27*	*198*	*213*
2. Based on Limited Area/Animals	14,309	19,275	23,000	22,574	24,544	24,451
D. Payments Based on Historical Entitlements (support programmes)	–	–	1,379	811	715	675
E. Payments Based on Input Use	6,913	7,871	7,847	8,333	8,682	8,812
1. Based on Use of Variable Inputs used	2,971	3,201	3,404	3,689	3,639	3,700
of which:						
Env. (2078/92), Conversion of Arable Land to Grassland and Forage	*61*	*66*	*96*	*250*	*255*	*276*
2. Based on Use of Farm Services	622	518	602	566	585	579
of which:						
Env. (2078/92), Training Demonstration	*46*	*16*	*24*	*46*	*58*	*62*
3. Based on Use of Fixed Inputs	3,316	4,152	3,841	4,078	4,458	4,533
of which :						
Env. (2078/92), Soil improvement	–	–	–	*108*	*52*	*56*
F. Payments based on input Constraints	2,118	2,117	2,616	3,807	4,062	3,722
1. Based on Constraints of Variable Inputs		3	53	46	328	261
2. Based on Constraints of Fixed Inputs used	2,062	2,068	2,119	3,189	3,119	2,796
of which :						
Env. (2078/92), Extensification of Livestock	–	*84*	*196*	*409*	*491*	*530*

Table 8.1
Continued

Policy Measures	EU—12		EU—15			
	1993	1994	1995	1996	1997	1998[1]
3. Based on Constraints on a Set of Inputs of which:	56	146	363	572	615	685
Env. (2078/92), Public Land, Leisure, Upkeep of Abandoned Land	*16*	*64*	*139*	*343*	*345*	*373*
G. Miscellaneous Payments	1,362	1,239	2,034	1,194	725	649
1. National Payments	389	265	1,962	1,109	635	553
2. Sub-National Payments	973	974	72	85	90	96
III. Percentage PSE	45%	42%	40%	34%	37%	45%
IV. Share of Regulation 2078/92 in PSE	0.2%	0.4%	0.8%	2.7%	2.7%	2.4%

Note: Above figures do not include support in the sense of general services such as research and development, agricultural schools, inspection services, infrastructure, marketing and promotion etc.
[1] Provisional estimate.
Source: Own calculations based on OECD PSE data.

tative approach. These policies were implemented in the eight STEWPOL countries. More specifically 14 were implemented in Belgium, 10 in the UK, 9 in Sweden, 30 in Germany, 14 in Austria, 3 in Greece, 15 in France and 18 in Italy.

The above policies were grouped into 9 general categories, according to their major objective (cf. Chapter 2), as follows:

1. Conservation of agricultural crops, rural landscape and related typical elements;
2. Natural and semi-natural environmental conservation;
3. Wildlife/Biodiversity;
4. Soil conservation and protection from erosion;
5. Recreation and access to agricultural land;
6. Reduction of negative impacts of agriculture;
7. Support to local production/quality labels;
8. Conversion of agricultural land into forests; and
9. Multi-objective.

The majority of policies were associated with categories 3, 6 and 9. As a next step, two methods were used for the analysis of results:

- the comparative statistical method
- hierarchical cluster analysis

Results provided by these two methods are reported below.

Table 8.2
Performance of policy-groups, all countries

Policy-Groups	Overall Score	Two Best- Performing Criteria	Two Worst-Performing Criteria
1	2.25	TRANSPARENCY, MONITORING	TARGETING, EVALUATION
2	2.15	TRANSPARENCY, MONITORING	EVALUATION, TARGETING
3	1.84	MONITORING, TRANSPARENCY	TARGETING, EVALUATION
4	2.12	MONITORING, TRANSPARENCY	TARGETING, TAILORING
5	2.45	TRANSPARENCY, MONITORING	TARGETING, EVALUATION
6	2.40	MONITORING, TRANSPARENCY	TARGETING, EVALUATION
7	2.38	TRANSPARENCY, MONITORING	EVALUATION, TARGETING
8	2.32	TRANSPARENCY, MONITORING	TAILORING, EVALUATION
9	2.16	MONITORING, TRANSPARENCY	TARGETING, TAILORING
Aver. Score	2.23	TRANSPARENCY, MONITORING	TARGETING, EVALUATION

8.5.2.1. Comparative statistical analysis

The aim of the comparative statistical analysis was to evaluate policies' adherence to the qualitative criteria and sub-criteria. In this regard an ordering scale of 1 to 5 was used. The closer the average score is to 1 and the smallest the standard deviation, the better the adherence to the criteria, or in better words, the better the criteria are met. Table 8.2. indicates the performance of agri-environmental policy groups in all eight STEWPOL states.

Regarding popular policies implemented, results indicate that policies associated with Wildlife/Biodiversity, were characterised by a very satisfactory performance (average score in all Member States, 1.84). Transparency and monitoring demonstrate high adherence, while the targeting criterion shows a low degree of adherence.

Policies whose objective was the Reduction of Negative Impacts of Farming in all countries, were judged to perform neither satisfactorily, nor unsatisfactorily (average score 2.4). Monitoring showed the best overall adherence (1.85), followed by transparency (2.08), while a low degree of adherence is associated with targeting.

The average score of multi-objective policies is regarded as satisfactory (2.16), with monitoring and transparency showing the best overall adherence, and targeting the lowest.

Other policies with a satisfactory level of adherence include Soil Conservation (2.12) and Natural and Semi-natural Environmental Conservation (2.15). In both cases, monitoring and transparency show the best adherence, while targeting is associated with a low adherence level. Finally, Recreation and Access to Agricultural Land (2.45) and Support to Local Production (2.38) show a low degrees of adherence, with transparency and monitoring being the two best-performing and targeting and evaluation the worst performing ones criteria.

In summary, the main points of the findings of the above analysis are:

- The best performing policy-groups are wildlife/biodiversity, soil conservation and protection from erosion, natural and semi-natural environmental conservation and multi-objective;
- The worst-performing policy-groups are recreation and access to agricultural land, and reduction of negative impacts of agriculture;
- The most adherent criteria are Transparency and Monitoring;
- The worst performing criteria are Targeting and Evaluation;

- The best-performing sub-criteria are Transparency-clear definition of type of instrument utilised and Monitoring-implementation of compliance of beneficiaries;
- The worst-performing sub-criteria are Transparency-direct link of payments to expected environmental outcome and Monitoring-cost effective implementation.

8.5.2.2. Cluster analysis

The objective of the cluster analysis carried out was to evaluate the agri-environmental measures' adherence to the qualitative criteria and also to specify homogeneous groups of policies. Homogeneous groups of the 113 policies were classified into 4 groups by applying the agglomerative hierarchical clustering method.

(a) *Cluster definition.* The responses to the hypotheses of the criteria have been transformed to z-scores and a *t*-test has been applied in order to assign the significance of the specific sub-criteria (hypotheses) in defining each cluster. For every cluster, negative z-scores for a specific sub-criterion imply that 'agree' responses which refer to a sub-criterion are predominant for policies belonging to this particular cluster as compared to all policies examined. Respectively, positive z-scores for a certain sub-criterion imply that 'disagree' responses are predominant.

Clusters are defined by the extreme responses associated with the specific sub-criteria, i.e. by the most negative (agree) and the most positive (disagree) z-scores. In other words, policies are distinguished in each different cluster according to the degree responses to the specific sub-criteria are homogeneous as well as the interaction of the responses.

The first cluster is characterised by the fact that it contains policies for which 'agree' responses are predominant (as compared to all policies examined) for the sub-criteria: (a) environmental outcome expected (A-TRANSPARENCY1) and (b) payments are properly calculated (A-TRANSPARENCY3). In addition, disagree responses are predominant (as compared to all policies examined) for the sub-criteria: (c) existence of evaluation procedures concerning the environmental output (D-EVALUATION1A), (d) existence of evaluation procedures concerning the economic efficiency of the measure (D-EVALUATION1B), (e) existence of monitoring procedures concerning compliance of the beneficiaries (E-MONITORING1) and (f) ensuring of compliance of the beneficiaries (E-MONITORING2).

In other words, for the policies of this cluster, payments are directly linked to the environmental outcome expected, while these payments are calculated for each type of product on the basis of the total area cultivated (or the number of animals) or the volume of production or inputs used. The absence of evaluation procedures concerning both the environmental output and the economic efficiency of the measure also characterise policies in this cluster. Finally, monitoring procedures concerning compliance of the beneficiaries with the terms of the contract are generally not implemented while according to these monitoring procedures implemented compliance of the beneficiaries is ensured.

Concerning the other clusters, all information provided in Table 8.3 should be interpreted in the same manner as it was done for cluster 1. It should be remembered that positive signs (+) correspond to responses associated with 'disagree' and vice-versa.

(b) *Results.* According to Table 8.4, there is a balanced distribution of the 113 policies, among the 4 defined clusters.

Turning to the detailed grouping of policies into the four clusters a certain degree of concentration of the policies of each country in one cluster can be detected. Thus, all but one of the Swedish policies are grouped in cluster 1, together with most policies of France. The Austrian policies are grouped (almost entirely) in cluster 2, those of Italy in cluster 3 and the vast majority

Table 8.3
Definition of clusters and statistical significance of responses by relevant sub-criteria (Z-scores)

Cluster 1	Cluster 2	Cluster 3	Cluster 4
A-TRANSPARENCY1 $t = 3.045$, Sig. $= 0.006$	A-TRANSPARENCY4 $t = 3.985$, Sig. $= 0.000$	A-TRANSPARENCY1 $t = 3.166$, Sig. $= 0.004$	B-TARG3 $t = 4.205$, Sig. $= 0.000$
A-TRANSPARENCY3 $t = 6.053$, Sig. $= 0.000$	B-TARGETING1 $t = 3.129$, Sig. $= 0.004$	A-TRANSPARENCY2 $t = 9.443$, Sig. $= 0.000$	B-TARG4 $t = 3.212$, Sig. $= 0.003$
D-EVALUATION1A $t = 3.376$, Sig. $= 0.002$	C-TAILORING2 $t = 4.159$, Sig. $= 0.000$	A-TRANSPARENCY3 $t = 3.218$, Sig. $= 0.003$	C-TAILORING1 $t = 3.760$, Sig. $= 0.001$
D-EVALUATION1B $t = 4.863$, Sig. $= 0.000$	D-EVALUATION3 $t = 4.070$, Sig. $= 0.000$	A-TRANSPARENCY4 $t = 3.502$, Sig. $= 0.002$	C-TAILORING2 $t = 8.209$, Sig. $= 0.000$
E-MONITORING1 $t = 6.131$, Sig. $= 0.000$	E-MONITORING1 $t = 3.447$, Sig. $= 0.002$	D-EVALUATION2 $t = 4.695$, Sig. $= 0.000$	D-EVALUATION1A $t = 5566$, Sig. $= 0.000$
E-MONITORING2 $t = 3.285$, Sig. $= 0.003$		D-EVALUATION3 $t = 3.161$, Sig. $= 0.004$	D-EVALUATION2 $t = 5.458$, Sig. $= 0.000$
		E-MONITORING2 $t = 3.163$, Sig. $= 0.004$	D-EVALUATION3 $t = 7.436$, Sig. $= 0.000$
		E-MONITORING3 $t = 4.304$, Sig. $= 0.000$	E-MONITORING1 $t = 7.761$, Sig. $= 0.000$
			E-MONITORING2 $t = 9.103$, Sig. $= 0.000$
			E-MONITORING3 $t = 3.696$, Sig. $= 0.001$

[1] 2-tailed significance at 97%

Table 8.4
Number of policies contained in each cluster

	No of Policies	% distribution
Cluster 1	25	22.1
Cluster 2	30	26.5
Cluster 3	28	24.8
Cluster 4	30	26.5
Total	**113**	**100.0**

of German policies in cluster 4 (with the remaining in cluster 2). On the other hand, the policies of Belgium are distributed in all four clusters, those of the UK in clusters 1, 2 and 3, and of Greece in clusters 2 and 3.

Cluster 1 mainly contains the policies of two countries (Sweden and France) with a very contradictory degree of adherence (Sweden is the second best-performing country, while France is the worst-performing one). Cluster 2 contains policies of six differently-performing countries, while cluster 3 is dominated by Italy (one of the best-performing countries), but there is also a presence of not so well-performing countries such as France and Belgium. Finally, cluster 4 is dominated by the German policies, which is the best-performing country in terms of adherence.

Turning to the results of the frequency analysis per country, the main findings are: Germany seems to be the country with the best adherence of agri-environmental policies, followed by Sweden and Italy. On the other hand, results are not satisfactory for France, Greece, Belgium and the UK.

In more detail, in the case of Germany, frequency analysis shows a very satisfactory adherence especially for the monitoring and evaluation criteria. Non-satisfactory adherence is detected only for sub-criteria A1 and B2. The performance of Sweden is also very high especially for the evaluation and targeting criteria. Non-satisfactory adherence is detected only for sub-criteria A1 and E3. Italy also performs generally well, especially in the case of the tailoring and transparency criteria; however, the existence of non-satisfactory adherence is detected, especially in the evaluation and targeting criteria, while the monitoring cost-effectiveness sub-criterion indicates zero adherence.

In the UK, adherence is satisfactory for the transparency criterion, while the tailoring and evaluation criteria perform badly. In Austria the situation is satisfactory for the transparency and monitoring criteria, but unfavourable in the case of evaluation, where 3 out of 4 sub-criteria score less than 50 per cent. In the case of Greece and France, only the transparency criterion performs satisfactorily, while there is a significant level of non-adherence in the remaining criteria. Finally, in the case of Belgium, there is satisfactory adherence in the monitoring and (secondly) tailoring criteria, but adherence is low in the case of evaluation and targeting.

Turning to the frequency analysis per defined cluster, the main findings are: The transparency criterion is satisfied (in general) in all four clusters, and especially in cluster 4, which mostly contains German policies and cluster 2 which contains Swedish policies. Problems of non-satisfactorily adherence are detected only for sub-criterion A1. The targeting criterion is satisfied in clusters 1 and 4, but not at all satisfied in cluster 2 and only partly in cluster 3 (where sub-criteria B3 and B4 fail).

The tailoring criterion is satisfied especially in cluster 4, but also in cluster 3, while results for sub-criterion C2 are problematic in the other two clusters. The evaluation criterion is satisfied totally in cluster 4 and partly in cluster 1, while results show a low degree of adherence in the case of cluster 2 and especially 3. Finally, monitoring is satisfied in both clusters 4 and 2, but problems with the cost-effectiveness sub-criterion are detected in the other two clusters.

8.6. Discussions and conclusions

The above analysis has used two complementary methodologies to assess the possible trade-distorting effects of agri-environmental policies implemented in eight EU Member States. Results have shown that in general the probable degree of trade distortion caused by the implementation of agri-environmental policy measures in the EU is rather insignificant.

Despite that the PSE suffers from several weaknesses (emphasis on transfers to producers and not on trade-distortions arising from agricultural policies; non-coverage of certain measures and policies financed outside the Ministries of Agriculture as well as those not considered as transfers to producers; calculation on a product-specific basis), it can surely draw some inference regarding the contribution of agri-environmental measures to total farm protection. This analysis has shown that the PSE share of agri-environmental policies in the percentage PSE has increased from 0.2 per cent in 1993 to 2.4 per cent in 1998. However, this share remains far below the critical WTO threshold of 5 per cent, which is considered as a limit below which trade distortion is considered as minimal or negligible.

Regarding the future, it is worth noticing that it is not certain that the implementation of the CAP reform could change the level of the percentage PSE, as the reduction of institutional prices is counterbalanced by direct income aid. Thus, much will depend on the evolution of world-prices.

The weaknesses that characterise the PSE, and give rise to doubts on its suitability as an indicator of trade distortion, led to the utilisation of a supplementary qualitative approach developed by the OECD. The strengths of this qualitative approach lay in the fact that it is measure-oriented, it evaluates whether payments are cost-effective in providing environmental benefits and at the same time non-distorting in agricultural markets, and can therefore assess eligibility of an agri-environmental measure under the 'green box' provisions of the AOA. Regarding the findings of the qualitative analysis, it has been derived that trade distortion caused by EU agri-environmental policies is rather low, while the situation might improve if more attention is attached to the criteria of targeting and tailoring. In general, policies demonstrate a satisfactory level of compatibility, however additional efforts are needed also with respect to cost-effective implementation and direct linkage to expected environmental outcomes.

Wildlife/Biodiversity policies are showing the best adherence to all criteria, an important finding given their wide application. Equally important is that policies such as Reduction of Negative Impacts from Agriculture and Recreation and Access to Agricultural Land show loose adherence, an important finding taking account the importance of these measures.

Comparison of the two methodologies emphasises their different output: PSEs are more closely indicating general distortions and the qualitative approach is more measure oriented. The latter has particular promise although further work is needed to obtain repeatable expert judgements.

References

Canhill, C. and Legg, W. (1990) *Estimation of Agricultural Assistance Using Producer and Consumer Subsidy Equivalents: Theory and Practice*. OECD Economic Studies No 13, Paris.

FAO (1973) *Agricultural Protection: Domestic Policies and International Trade*. Conf. Doc., c73/Lim/9, Rome.

FAO (1975) *Agricultural Protection and Stabilization Policies: A Framework of Measurement in the Context of Agricultural Adjustment*. Conf. Doc., c75/Lim/2, Rome.

FAO (1995) *The measurement of the impact of environmental regulations on trade*. Committee of Commodity Problems, FAO, Rome.

Goldin, I. and van de Mensbrugghe, D. (1995) *The Uruguay Round: an assessment of economywide and agricultural reform*. World Bank Conference on 'The Uruguay Round and the Developing Economies', January 1995.

Harrison, G., Rutherford, T. and Tarr, D. (1995) *Quantifying the Uruguay Round*. World Bank Conference on 'The Uruguay Round and the Developing Economies', January 1995.

Hertel, T. W. (1989) PSEs and the Mix of Measures to Support Farm Incomes. *The World Economy*, March, 1989.

Krissof, B., Ballenger, N., Dunmore, J. and Gray, D. (1996) *Exploring linkages among agriculture, trade and environment: issues for the next century*. Environmental Division of Agriculture, Agricultural Economics Report No 738.

OECD (1987) *National Policies and Agricultural Trade*, Paris.

OECD (1995) *The Uruguay Round: A Preliminary Evaluation of the Impacts of the Agreement on Agriculture in the OECD Countries*, Paris.

OECD (1997) *Agricultural Trade and the Environment: Anticipating the Policy Challenges*. Joint Working Party of the Committee for Agriculture and the Environmental Policy Committee OECD, Paris.

OECD (1998) *Agriculture and the environment: issues and policies*. OECD, Paris.

Pearce, D. and Turner, R. (1990) *Economics of Natural Resources and the Environment*. Baltimore, The John Hopkins University Press.

Tobey J. (1991) The Effects of the Environmental Policy Towards Agriculture on Trade Some Considerations, *Food Policy*, 16(2), 90–93.

Tobey J. and Smets H. (1996) The Polluter Pays Principle in the Context of Agriculture and the Environment. *The World Economy*. Vol. 194.

G. Van Huylenbroeck and M. Whitby
Countryside Stewardship: Farmers, Policies and Markets
© 1999 Elsevier Science Ltd. All rights reserved

Chapter 9

Conclusions and policy recommendations

Guido Van Huylenbroeck and Martin Whitby

Abstract—In this concluding chapter the main findings from the previous chapters are summarised and further discussed in the light of future countryside policy recommendations. Based on some theoretical reflections and the results reported in the previous chapters, the actual implementation of agri-environmental schemes as well as the possibilities for improving the efficiency of these policies are discussed. Questions about the desirability of these policies and possibilities for cost minimisation and how to better target them towards environmental problems are raised. Conclusions as to the long-term contribution of such policies are also formulated.

9.1. Introduction

In previous chapters, CSPs applied throughout the eight STEWPOL countries have been analysed in several ways. The aim of this chapter is to provide an overview of major conclusions and to formulate some observations and recommendations with regard to the future implementation of these policies. This is certainly relevant as the EU, in its Agenda 2000 reform of the CAP, has clearly indicated its intention to continue along the road of agri-environmental policies and to provide extra payments for farmers in eligible areas for positive contributions towards the environment. Before the end of 1999, Member States have to make proposals for what might be called 'second generation' CSPs. There is no doubt that these policies and the payments provided will also form an important issue in the next WTO negotiation round, starting now, where further steps towards liberalisation of agricultural trade will be discussed. Agri-environmental and rural development policy, as part of structural policy, are also regarded as a way to smoothly integrate new Member States into the EU.

The chapter is organised as follows: first the main outcomes of the research project are listed, then some discussion is given on the following points which are discussed in turn: objectives of CSPs and their realisation, the costs of CSPs and how to decrease them, output and trade effects and their implications and the conformity of policies with general policy principles raising the question of acceptability. Finally some conclusions with respect to future application of CSPs are formulated. However, in reading this chapter, it should not be forgotten, as already indicated in Chapter 1, that agri-environmental policies cannot be evaluated in isolation from the rest of CAP-policies of which they form only a minor part with a small impact compared to other policies.

9.2. Main outcomes of the research

Without being exhaustive, the following results are derived from the theoretical and empirical work reported in previous chapters. They are presented under the three key words of this book: policies, farmers and market effects.

9.2.1. Policies

- Both agricultural practices (e.g. fertilisation, mowing, etc.) and countryside stewardship practices (water management, maintenance of landscape elements, etc.) are inputs to the 'rural fabric', jointly generating multiple outputs (food and fibre, environmental and recreational goods, landscape, biodiversity and others);
- The majority of agri-environmental policies applied tries to stimulate countryside stewardship behaviour of farmers through changing prevailing economic incentives between F&F production and the production of EGSs. Although rather traditional regulating measures are used such as output control measures (limits to cattle density), input control measures (ban or limitation of fertiliser, pesticides or other contaminants), production premiums (payments for hedges, stone walls and so on) or combinations of them, what is different about CSPs is that the majority of them involve individual contracts between farmers and the State;
- Major objectives of the policies are the reduction of negative impacts of agricultural practices (mainly in order to protect surface and ground water quality) and the conservation of landscape, environment and wildlife. The outcome of most policies are public goods although the link with agricultural practices allows their marketing as joint products in certain cases;
- Most policies have multiple, rather loose and ill-defined objectives and do not refer to reference points or specific targets, making the evaluation of benefits difficult;
- Based on a set of indicators reflecting environmental objectives, economic efficiency, cost of policies, scope and uptake, enforceability, persistency and compatibility with general policy principles, seven distinct groups have been identified among the measures applied. Groups can be characterised by the level of input constraint and by the output character of the EGSs produced (see Chapter 3);
- All forms of policy intervention, irrespective of their form, impose varying levels of administrative costs on the system. Countryside stewardship policies are characterised by high implementation and transaction costs in comparison with market regulation policies. These transaction costs have to be taken into account when comparing policies as they constitute the social opportunity costs;
- Based on the cost assessment made for case study schemes in Chapter 4, a rough estimation indicates that these transaction costs amounts up to 2 billion Euro across the EU for Regulation 2078/92 schemes since they were first introduced, which is not negligible compared to the total spending of the EU to farmers' compensations under this regulation of 7 billion Euro. Even if it is assumed that Member States also contribute up to 50% in direct costs, the transaction costs represent some 20% of total expenditure or up to a 40% addition to local expenditure;

– Analysis of the transaction costs of policies in different countries reveal striking differences indicating the importance of administrative structures for transaction costs as well as the existence of scope to decrease them.

9.2.2. Farmers

– Uptake by farmers depends in the first place on the economic incentive offered: the higher the compensatory payment, the greater the farmers' willingness-to-accept the required constraints of CSPs. The hypothesis that farmers' uptake is positively correlated with the level of compensation and their own direct utility derived from EGS-production, and negatively with the direct costs as well as the private transaction costs connected to EGS-provision, is confirmed by the analysis of the farm survey results;
– Besides compensation, uptake is higher among better educated and informed farmers and among those with a positive attitude towards the environment. This indicates that education, information and extension matters and can gradually shift farmers' opinion about agri-environmental policies;
– A majority of participating farmers surveyed would continue, be it to a reduced extent, always the required conservation efforts after the end of their contract. This can be explained by the fact that in a number of cases farmers are paid to maintain the actual situation in order to prevent intensification (see also Chapter 2) and that policies help in changing farmers' attitudes through learning by doing in adopting more environmentally friendly technologies;
– Although country differences can be observed, only a minority of farmers indicated that the schemes decrease the overall output and profits of their farm. This may be explained by the short period since implementation, whereas input restrictions only generate output effects after some years.

9.2.3. Market effects

– With regard to the effects on agricultural commodity output, a distinction has to be made between policies aiming at restoring landscape elements, policies designed to keep farmers on the land in marginal areas and policies limiting the use of inputs or constraining the output level. For the first two categories it is clear that the output effects are non-existent or limited. In the long run they can even have the effect of reducing the impact of market regulation policies by keeping production factors in agriculture. But also for the last category of policies the analyses in Chapters 6 and 7 show that the total net output effects are in general small, even when applied on a large scale as in Austria. One possible explanation can be found in shifts in the production plan and in the fact that (part of) the output reduction effect is offset by an output increasing effect as part of the compensatory payments are used to buy additional, non-restricted inputs;
– Significant output reductions as a consequence of CSPs can only be expected when input reducing programs are introduced in favourable areas on a large scale;
– The modelling exercise also indicates that in general the income effect for participating farmers is positive, raising the question of over-compensation and rent creation (mainly due to flat rate payment systems);
– Calculation of shadow prices for land indicate that CSPs have an increasing effect on land values and thus probably also on land prices, though probably less than other CAP instruments;

- The analysis of the Producer Support Estimates (PSE) of the application of agri-environmental measures in the EU reveals an increase of the PSE share of agri-environmental policies (due to higher share of the EU budget allocated to them) although the actual share is still far less than 5% of total PSE;
- The suitability of PSE as an indicator of trade distortion for agri-environmental measures can be challenged. A second, more qualitative examination has analysed the conformity of policies with the criteria OECD uses as indicators of non-trade distortion (transparency, targeting, tailoring, evaluation and monitoring) showing that in general the policies applied satisfy to a large extent the non-distortion criteria. Lowest scores were obtained for the targeting and tailoring criteria indicating that the correspondence with the intended environmental outcome seems in a number of cases rather loose. Also the evaluation criterion is a problem for a number of policies, probably because of the long time period before certain environmental benefits can be observed and the difficulty of finding good environmental indicators. But in general the analysis indicates that the likely degree of trade distortion caused by the present EU CSPs is insignificant.

9.3. Discussion

9.3.1. Are countryside stewardship policies an efficient way to achieve countryside management objectives?

As indicated in the Chapter 1, the major objective of agri-environmental policy intervention is to change farming conditions in such a way that farmers are (economically) encouraged to take countryside management and stewardship constraints into account in the management of their farm and land. This is necessary because landscapes and biodiversity in rural areas are public goods which receive little or no encouragement through markets.

The central question is whether remuneration of farmers in exchange for a limitation of inputs or outputs is the most efficient way to reach this goal as compared with other instruments. The answer is complex and difficult and is partly beyond the scope of this book. Theoretically, payments are only justified if practices or services are requested which farmers do not apply under non-distorted conditions and for which they are not compensated through the sale of their products. An example of such a non-remunerated and usually not applied practice can be the development and maintenance of a buffer strip around the fields. The analysis in Chapter 2 has shown that much CSPs remunerate farmers for traditional practices which would be normal under non-distorted market situations or under more severe environmental standards. A strong interpretation of the Polluter-pays-principle (PPP) would preclude such compensations. However as long as both input and output market prices are distorted and positive and negative externalities are not fully reflected by the prices of the commodities produced, CSPs may be regarded as a second best solution.

Market distortions both on the input and on the output side also have important consequences for the compensatory payments and prescriptions of CSPs:

- the remuneration has to be higher because foregone incomes are higher than in non-distorted markets and because the prices that farmers receive for environmentally friendly produce are not high enough due to surplus production of conventional products;
- payments now often have to counteract negative effects of market distortions (cf. the example given in Chapter 6 for the beef extensification premiums);
- the more strongly environmental pollution is regulated, the lower the need for payments above regulatory standards.

Therefore CSPs cannot be judged in a vacuum but must be assessed within an existing policy environment. Many of them can be regarded as a kind of transitory instruments necessary to protect or enhance environmental values at risk because of existing price distortions in the market. However even if price distortions were to be removed, certain forms of CSP may still be justified to counteract remaining external effects and improve the allocation of resources.

The distinction between positive and negative externalities is a matter of benchmarks and in certain sense (legal) conventions. This may raise, as discussed in Chapter 2, questions about internal conflicts between environmental attributes or about the time reference. What is now regarded as a positive externality (for example the planting of hedges or a decrease of fertilisation in order to increase biodiversity) may well be the consequence of negative externalities in the past (such as the removal of existing hedges or intensification of fertiliser input that has decreased biodiversity). Taking the actual situation as a benchmark when the policy objective is to conserve an existing situation or, for second generation contracts, to maintain the improvements produced by the first generation contract, would be contradictory or even wrong when calculating compensatory payments based on the income foregone principle, because incomes remain the same. Fixation of reference points is on the one hand crucial to defend countryside stewardship policies (certainly in an international forum) but on the other hand it is not an easy task, and requires careful judgement and recognition of the dynamics and benefits of the agri-environment.

Although no valuation of environmental benefits has been made, the typology of policies gives insight into the values that are produced, while narrowing the efficiency question down to the differences between CSP instruments. In general, environmental outcomes per hectare will be higher for policies banning or greatly reducing the use of fertilisers, pesticides and other contaminants (policies belonging to group 1 and 3 in terms of the classification given in Chapter 3). Other policies also providing high levels of benefit, be it of another nature, are those policies of which the objective is to conserve landscape elements, cultural heritage or specific ecosystems (the policies belonging to group 5 and partially to group 7). Although the data used in Chapter 3 are not totally precise, the policies with high estimated benefits seem also to be the most expensive ones (requiring high compensations to farmers and/or transaction costs), indicating the existence of a trade-off between environmental benefits and exchequer costs. Broader policies of the "*prime à l'herbe*" genre are cheaper but their outcomes are more uncertain as they mainly try to avoid the disappearance of existing amenities at risk from distorted market conditions. However, this may be a justifiable objective as long as these market distortions exist.

9.3.2. Improving the outcome of policies

To increase the outcome of CSPs, policy design offers three main possibilities:

– targeting of policies through specification of eligibility criteria and zones in which the policy is applied and more focused prescriptions;
– increasing uptake by modulation of compensation levels and better extension programmes;
– improvement of policy administration and mechanisms.

9.3.2.1. *Targeting of policies*

The aim of better targeting policies is twofold: on the one hand avoiding dead-weight losses and on the other hand making policies better coincide with the problems identified and improving outcomes. Better targeting of policies can be reached by:

- zoning;
- specification of eligibility criteria;
- specification of prescriptions.

Zoning is primarily used to allow differentiated payments and to better link the prescriptions or requirements with problems encountered. The nature of the problems must form the basis of the zoning and not only administrative boundaries. Designation should be based on ecological, agronomic and socio-economic criteria (cf. the English ESAs).

Better targeting can also be reached by specifying the eligibility criteria for farmers who own or use land with specific features that need to be protected or enhanced. This last strategy can easily be followed for landscape or cultural heritage elements, but is less evident for other environmental values. Policy implementation can also focus on certain socio-professional categories such as professional farmers, part-time farmers, agricultural companies and co-operatives although these subdivisions can be questioned when considering multi-functional agriculture.

Another way to target the agri-environmental payments better is through tighter specification of the prescriptions, in particular when there is a strong relation between an EGS and applied practices. It is open for discussion whether the requirements must prescribe the use of a certain technology or focus on environmental outcomes, leaving the choice of technology to farmers, assuming a development towards more sustainable technologies. The empirical evidence from the inventory used in Chapter 3 clearly indicates that policies enhancing the use of more sustainable technologies, such as mechanical weeding instead of using pesticides, are in the long run less reversible and thus more persistent.

Research on more sustainable technologies should be stimulated as persistence of effects after the removal of payments is an aspect which deserves more attention. Although it may be more expensive in the short run, priority should be given to policies of which the benefits are more likely to be continued than of those only temporarily renting the property rights. A related issue is the optimal contract length. Perhaps longer contracts imply higher payments, but they usually extend the sustainability of the provision of EGSs, in particular when very specific, or in the language of Williamson (1985) idiosyncratic, investments are involved.

A major discussion point related to the effectiveness of policies is their voluntary nature. Given the non-separability of the provision of some types of agri-environmental public goods, such as landscape, the effectiveness and efficiency of voluntary measures can be doubted, because to make all farmers comply with the protection rules, very high compensation levels are required to convince the last farmer. This in effect passes rents to other farmers. By making the prescriptions mandatory (e.g. through designation of special interest areas for which specific exploitation rules are enforced) policies can be made more effective and efficient. This implies a partial or total shift of the property rights to the State (see also Bromley, 1991). A problem is that linking compensatory payments to mandatory prescriptions in designated areas, to make land exploitation under the rules imposed sufficiently attractive for farmers, may not be compatible with the PPP. Given these objections, a possible route to progress might be the delivery of environmental goods through collective approaches. Further research is required to know how such arrangements might fit into the existing framework based on individual management agreements and what the transactional cost implications would be.

9.3.2.2. Increasing uptake

To be effective both from an environmental point of view (see 9.3.2.1) and for commodity output control, uptake is a decisive element. The empirical evidence both from the Chapter 3 (the typology) and from Chapter 5 (the farm survey), clearly indicates that higher compensatory payments result in higher uptake. If society wants a public service from farmers it must be prepared to pay for it. Average reported compensations fall within an interval between 50 and 300 Euro per hectare with an average around 200 Euro per hectare. The payments have to ensure the compliance with the technical standards (compensations) and also incite farmers to participate (incentives). To ensure high uptake these payments should also cover private transaction costs unless society prefers to depend on farmers' altruism.

Incentives and conservation payments, meaning remuneration for the maintenance of an existing situation, are sometimes criticised as being over-compensations, but may in the long run be necessary to avoid further degradation of the environment or in more marginal areas to keep farmers on their land in order to ensure the conservation of typical landscapes. In this sense, the dynamic context of agri-environmental policy needs consideration: for example, in terms of attitudinal change over time, experience of scheme interactions on both sides (farmers and state agencies), and the build-up of human capital (and of environmental capital), necessitate a long-term view on participation for efficiency. This is already recognised by Regulation 746/96 which allows for 'surplus' incentives over income foregone. If society wishes to ensure certain services in the long run, it has to accept that this creates certain rents for those offering this service as capital and labour are also rent seeking inputs when the objective is environmental protection. It is of course also the role of policy design to find ways to reduce the rents society has to pay and to avoid these rents offsetting the objectives of agri-environmental policies.

The empirical evidence further shows that payments are in general lower the more farmers are remunerated according to their level of stewardship. This implies differentiated payments according to local conditions, differences in yield or livestock grazing density. This can be achieved through better targeting of policies (see above) or through individual negotiated agreements, both, however, resulting in higher transactions costs.

Another important aspect when seeking to increase uptake is information and education. The farmer survey indicates that there is still a problem of asymmetry of information between farmers and public agencies. Actions promoting the schemes should balance this asymmetry: experimentation with sustainable practices, demonstrations, campaigns of promotion and so on, but also general education about the importance of countryside stewardship are needed to make farmers more aware of their role and create a positive image around these activities. The survey results strongly indicate that better informed farmers and those with a more positive attitude towards environmental issues have a higher acceptance rate. Emphasising the role of farmers as stewards of the land and natural resources is an important issue for farmer training which probably remains too much production-oriented.

9.3.2.3. Policy administration and mechanisms

Environmental outcomes can also be ensured by better administration and implementation of policies. As already discussed above, the environmental benefits are higher when policies are adapted to local circumstances or needs. This implies vertical policies implemented and administered at the local level as local administrations are better aware of local needs, with the role of regional, national and EU authorities restricted to the conception of a general framework, financing and control. On the other hand the empirical evidence clearly shows that local administered policies impose higher transaction costs than more general horizontal policies. The importance therefore is not whether a policy is vertical or horizontal, but whether it is implemented at the right level and whether the right policy mechanism is used. Theoretically different possibilities exist which are not yet applied in practice (cf. Chapter 2 as well as the discussion in Hofreither (1998); Latacz-Lohman (1998) or Weaver (1998)). Further search for creative but applicable mechanisms should be stimulated.

For more general objectives, like conservation of grasslands or wetlands a more horizontal approach may be appropriate, while for biodiversity objectives or protection of specific landscape features a more locally administrated measure will be more efficient. Of course the administrative structure and tradition of countries is a major aspect in this. Depending on the degree of state decentralisation, countryside stewardship is in general the competence of regional authorities (see Chapter 1). But even if this competence is fixed by countries' constitutions, the decision-maker can assign the execution of the various tasks (implementation, application procedure and control) to either central or decentralised public agencies or even hire services from the private sector. This institutional choice is of great importance, also for the success of a policy in terms of uptake, and should therefore be based on a careful analysis of costs and effects raising questions such as: who is the least-cost provider of any given transactional service (farm mapping or farm conservation planning, for example), what effect would the use of private organisations have and how can the quality of these services be assured.

Another (related) aspect, which will gain importance in the light of making agri-environmental policies more acceptable at the WTO negotiating table, is the monitoring, evaluation, control and enforcement of policies. One problem with regard to OECD criteria for non-trade distortion (Chapter 8), is the low score for evaluation. It is clear that environmental effectiveness increases with better control and monitoring. Efforts should be made to find better evaluation and monitoring systems. The fact that the inventory indicates that control of a number of policies is rather difficult and/or not very reliable is an important indication. The type of control depends of course on the nature of the CS output, the requested modifications of farming techniques or the provision of observable externalities. The decision-maker can decide on the degree and the frequency of the controls as well as on the nature and level of the sanctions. A combination between frequency of control and severity of sanctions must be found to ensure compliance and to minimise control costs. This can conflict with a general accepted legal rule that penalties should not be higher than the eventual advantage somebody gained from a policy. In this sense cross compliance with other payments such as direct income support may not work.

A possibility of saving on control costs, which brings us to the next section, is a better integration between several administrations. At this moment agri-environmental payments are still too much regarded as separate from other agricultural policy administration, while often the same information is required. A reorganisation of administration and control procedures could probably increase efficiency and decrease administrative costs (cf. the carrefour idea already discussed in the EU as a means of improving rural development (EC, 1990).

9.3.3. The transaction costs of policies

As indicated in Chapter 4, the total costs of CSPs can be divided into two parts: on the one hand the compensation given to farmers or land users and on the other hand the running cost of policies, often referred to as transaction costs. The sum of the two is the expenditures society must make to ensure the provision of a certain benefit, in this case the environmental benefits generated by the policies. However, from an economic point of view, it is important to make the distinction as remuneration payments are only transfers, where transaction costs represent real social opportunity costs. It is therefore essential to include them in economic analyses as public provision should only be considered when total costs are lower than under market mechanisms or when market transactions desired by society do not take place.

With respect to the payments to farmers, information is easily available. This is not the case with the transaction costs which are therefore (Chapter 4) called the 'invisible' costs of agri-environmental policies. Indeed, the running of these policies requires important expenses to install, monitor and control the envisaged actions, but these costs are normally not reported in public expenditure budgets.

The information collected through the STEWPOL-project and reported in Chapter 4 indicates the importance of these costs and shows that they are by no means negligible and are higher than the transaction costs of other policy instruments. This says nothing about policies being justified or not, as for that a comparison with the benefits is required. But even without this comparison, analysis of transaction costs of public policies is of interest to detect inefficiencies in management and design of policies and to see if their existence or amount can be explained, justified or reduced.

A first observation is the high level of transaction costs during the initiation of policies, due to the particular costs of setting them up. Over time the ratio of administrative costs to compensation payments tends to fall. This may be evidence of a learning effect making running costs of policies decrease when administrators gain experience with it. In this aspect the reported costs are probably higher than they will be in future because now all countries are acquainted with these policies and have organised their administration accordingly and because of economies of scale as familiar policies applied more widely. In this sense the marginal or incremental cost of an extra agreement is more important than the total administrative cost as it seems to be far more expensive to establish management agreements than to maintain them, so set-up costs should be seen as an investment, and their returns maximised once they are made.

In terms of policy appraisal, a key question is whether the cost-effectiveness rankings of schemes would change if their transaction costs were included. Although more empirical evidence than that reported in this book is necessary to answer this question, the evidence reported here does indicate striking differences between policies, depending on the administrative structure in which they are embedded, the level of administration, contract length, the prescriptions to be respected and so on. A careful analysis of administrative costs is thus required to see how far transaction costs reduce the cost-effectiveness of policies.

However, such analysis must be related to the extent of achievement of policy objectives. For a more valid assessment of cost-effectiveness of agri-environmental schemes, there is a need to examine the benefit side of agri-environmental policies and then set the marginal costs of schemes, including their organisational costs, against marginal benefits. Such a benefit analysis not only requires more evidence from ecologists or environmentalists, but also raises the need of assessing the public appreciation of benefits. Until now only fragmented knowledge is available. The development of a common and internationally accepted procedure for benefit assessment, although demanding considerable resources, could be a big step forward. Both the systematic use of benefit assessment methods in the US as well as the handbook of the International Association of Water Studies (1996) for evaluating water conservation or improvement programmes, show that such benefit assessments are possible, allowing more complete evaluation of programmes or measures. More systematic research is urgently needed.

A more contentious question raised in Chapter 4 is who has to pay the transaction costs? In other words, is it defensible that part of the administrative costs should also be eligible for Community funding, in order to avoid the potential bias that Member States currently have an incentive to favour lower transaction cost but higher compensation cost schemes? The present regime tends to support more general, but from an environmental point of view less effective, schemes rather than closely-targeted but more effective ones. So there is a potential risk of developing policy frameworks that are inefficient in terms of overall value for money.

The trade-off between organisational and compensatory expenditures is therefore crucial and needs to be better taken into account in future policy design, in particular because an extended use of instruments such as management agreements can be expected, given the options in Agenda

2000. Theoretically compensatory payments can be minimised by differentiation of payments according to zones or individual negotiations as this avoids the rents that are paid when flat standard payments are used. But as already indicated such differentiation implies higher transaction costs. So a social optimum minimising total costs must be found implying the search for an optimal degree of regional differentiation, mandatory character and individual negotiations (cf. the discussion above about the most appropriate administrative level).

9.3.4. Market and trade effects

Another major research point of the STEWPOL project was the effect of CSPs on output of agricultural commodities. For most policies the direct effect is a decrease of the output per hectare of agricultural commodities, at least as far as they require limitation or banning of external inputs. However, as indicated by the inventory an important number of policies do not affect output as they pay specifically for conservation efforts (such as maintenance of landscape elements) or seek to conserve threatened farm practices. But even for those policies where agricultural output is decreasing per hectare, it does not unambiguously mean that these policies have an overall output reducing effect because the compensations received by farmers can keep resources on farms which would otherwise leave the sector. They may also be used to intensify production on other parts of the farm.

Several policies are explicitly aimed at preventing land abandonment. The rationale for such policies is to promote the provision of landscape amenities that are linked to agricultural production. This environmentally motivated goal can clearly only be reached if agricultural output is produced. The question is whether such policies actually increase overall output or not. The results from the regional models in Chapter 6 show that in general the 'no CSP' scenarios do not result in significant higher outputs for the commodities in the model and therefore such policies can be considered to be output neutral at present.

Therefore agri-environmental policies are not the tools of first choice if the goal is to reduce farm output significantly because:

- CSPs are in general more expensive, certainly when transaction costs are taken into account, than conventional attempts at output control to induce the same output reduction;
- CSP measures provide payments that, if not applicable to the whole farm or area, may fund an option to intensify other parts of the farm;
- An important objective of agri-environmental schemes is conservation of existing practices or preventing land abandonment, implying slight output reducing effects at least in comparison with the 'policy-off' situation.

The output effect may also not be totally neglected as a small reduction such as estimated by the models (2–5 %) can be important at the margin in a surplus situation. Even if the net output effect is nil at least these policies limit further increases of agricultural production. The model calculations reported in Chapter 6 also indicate that with a reduction of price support, as in Agenda 2000, a slight increase in uptake can be expected, but not sufficient to really affect overall output of agricultural commodities. To obtain a measurable effect on output more general CSP schemes need to be offered with higher payments to attract also farmers in favourable zones, where larger output effects would be expected from extensification. It is nevertheless debatable whether this is the best solution from a social welfare point of view, because this would imply wider application of more soft and generic policies, which as indicated by Chapter 3, are less effective from an environmental point of view and, due to their wider scope, very expensive in total.

However, for targeting specific problems, such as conservation of farming activities in remote areas, agri-environmental measures may be better than general support instruments, at least under actual price conditions, as they are more focused. The shadow prices for land generated by the models in Chapter 6 indicate that these measures are highly effective at keeping land in production.

A larger application of agri-environmental policies will increase the share of support going to farmers through these systems, which could be regarded by non-EU countries as trade distorting. Given the increased possibilities in Agenda 2000 a related question is then how far CSP compensations are in line with requirements about non-trade distortion as put forward by WTO, and to what extent they will remain acceptable as belonging to the Green Box. Then the nature of the payments becomes more important. Although the analysis in Chapter 8 indicates that the policies satisfy the non-distortion criteria of OECD to a large extent, the low scores for the targeting and tailoring criteria indicate that the correspondence with the intended environmental outcome must be improved as well as the evaluation of policies. The EU has thus an interest in improving these aspects.

9.3.5. Political acceptability

The above discussion brings us to the point of the consistency of the policies with general policy principles and thus their acceptability from the point of view of fairness. The following aspects were investigated and will be briefly discussed:

– Who has to pay the compensation or incentives: the taxpayer through the state-pays-principle or the consumer through applying the beneficiary-pays-principle?
– To what extent do agri-environmental policies respect the polluter-pays-principle?
– Who gains and who loses from these policies?

With regard to the first aspect, it is clear that if possible the beneficiary should pay for the provision of goods he or she enjoys. In practice, as the output of countryside stewardship are mainly public goods, this is not easy to organise and only possible if the public good can be transformed either into a quasi-private good or into a club good, by assigning property rights to an individual or a collective or can be tied to a marketable joint product. If not, a collective payment mechanism must be used. The next option then is the State-Pays-Mechanism meaning that at the end it is the taxpayer who pays for the provision of the goods, whether or not with Common Funding. Part of the financial burden can also be placed on farmers by requiring certain services without extra payment (for example through cross-compliance) or by a levy system. Partial application of the farmer-pays-mechanisms can be justified from the point of view of the PPP (in its weak version only in cases where farmers do not comply with legal standards; in its strong version in all cases where the marginal benefits for producers are higher than the marginal costs of pollution abatement).

Another possible approach, at present only applied to a limited extent, is a compromise between the State- and the Beneficiary-Pays-Approach and consists in the development of collective funds through which those interested in the provision of the EGS can contribute to their payment. This form is applied on a small scale by trusts or other environmental organisations. Wider application might be that part of the price of certain food products (which correspond with a certain farm practice) is passed through a collective fund that can be used to finance individual or collective countryside management actions (cf. the extra charge for users of certain credit cards for charitable purposes).

The first option, the 'beneficiary-pays-principle' is already applied in a number of cases. The policies belonging to what has identified in Chapter 3 as the G2-policies (organic and labelled production support) are based on this logic. There are also a few cases where farmers are charging for access to their land for recreational purposes. As already indicated in Chapter 3 a problem is that the link between prescriptions and expected environmental outcome is not explicitly defined, making it both uncertain and difficult to sell and to defend support that goes further than providing a conversion incentive. A second problem of such policies is their compatibility with international trade principles. We do not refer here only to green or blue box measures, but to the question of whether the labelling of products is a protective measure against non-labelled produce.

For such policies the State has a role in the assignment of property rights as this is a precondition for market creation. However, its role should be limited to establishing conditions to create markets without paying for their creation itself at least when no other clear objectives can be demonstrated. Hence, once established, the market should remunerate the producer's efforts. In this respect the difference between quasi-private goods, where individual property rights are assigned to certain public goods, and so-called club goods (Hodge, 1991), where a collective property right is established, should be further investigated. In particular the last approach, based on collective action could gain importance in the future.

The PPP is another strongly debated issue. As indicated by the empirical results, for a number of policies there is doubt about their compatibility with this principle. Legally the PPP is only violated when farmers are paid for efforts which are legally required (weak version). However, the absence of clear thresholds ensure that in a number of cases, the PPP is not really applied. As already indicated in Chapter 2, formulation of clear thresholds is not easy and different aspects such as the physical environment, the nature of the environmental goods (because there can also be conflicts between them) and the use of the land should be considered. Indeed the PPP-discussion is in a sense a side issue as the real question is the position of the equilibrium point between the (actual and future) utility curve capturing consumer preferences and the production possibility curve between the production of F&F and production of ERGs and with which instrument this equilibrium point can be reached. Also transaction costs must be taken into this calculus as taxes or levies may also impose important administrative costs (depending on the payment vehicle).

It is also appropriate to recognise, if only briefly, the distributional aspects of CSPs. Where such policies discriminate in favour of larger farmers (as both output and area payments almost always do) there may be a strong case for modulating them. Sensibly applied, such modulation, for example by targeting more marginal farms, can introduce distributional justice into policy systems and greatly reduce policy costs. At least these issues should be recognised as an unintended policy outcome which in practice should be corrected through the taxation system. Another effect worth mentioning is that a number of CSPs are labour intensive and thus create jobs in the rural sector. The distributional aspect of transactions costs is also worth noting in that their very existence generates income and work for other than farmers (administrators) which otherwise would not exist. Also, among the beneficiaries from CSPs are those who enjoy the amenities provided. Most of these beneficiaries are from the middle class, who might claim that they also pay for this pleasure through their taxes.

9.4. Conclusions

The STEWPOL research project has tried to analyse and evaluate the actual application of agri-environmental and countryside management policies in 8 EU countries. The project had three important focal points:

(a)   the study of key attributes of the policies applied;

(b)   the perception of these policies by European farmers and their motives for participation or non-participation;
(c)   a normative study of the effects of these policies in controlling output.

The first point was studied by collecting information from legal and official documents and discussions with policy administrators. Attributes such as the number of participants, the area, the amount of compensation, the objectives, the required practices and the adherence to a number of policy indicators, have been inventoried and statistically analysed in the Chapters 2, 3, 4 and 8. Farmers' opinions have been surveyed and discussed in Chapter 5 while the potential output effects have been analysed through regional mathematical models in Chapter 6 and with an equilibrium displacement model in Chapter 7. The major conclusions have been listed in Section 9.2.1 of this chapter and have been further discussed.

Without being exhaustive and referring also to the conclusions of the other chapters a number of recommendations with regard to future application of CSPs is formulated in Box 9.1.

As a conclusion, the evidence so far confirms that both agricultural and stewardship practices are necessary to obtain rural environmental goods and services. The countryside stewardship factor has been neglected by farmers because of changing technological possibilities and economic conditions. Therefore the main role of agri-environmental policies is to change the production conditions for farmers in favour of landscape management and conservation efforts so that they will again pay more attention to it. Payments have the role of translating to farmers the stated and revealed preferences of 'consumers' with regard to the provision of public goods. The farmers' remuneration can either be direct, through compensatory payments financed by the taxpayer or be indirect through lower production costs or extra market benefits.

Agri-environmental policies in the EU are a first and recently introduced attempt in this direction, but income transfers remain of minor importance when compared to other instruments. In general these attempts seems to be successful. The following points can be suggested to further improve the efficiency and effectiveness of policies:

– more explicitly indicate thresholds and objectives in order to ensure payment for real additional contributions;
– better targeting policies by an optimal mix of zoning, eligibility criteria and required practices;
– focus more on long-term effects after removal of payments and on how sustainability of benefits may be attained;
– include transaction costs from the very beginning of the design of policies in order to avoid mis-allocation of public money;
– work for more integration and coherence between different policies;
– introduce stronger evaluation procedures for assessing benefits of policies and their conformity with international requirements and general policy principles, such as removal of all incentives to pollute.

Under these conditions, agri-environmental policies can play a more prominent role, certainly in the transition period to less government intervention on agricultural commodity markets. The most important conclusion is that agri-environmental policies should be used to stimulate farmers to deliver countryside stewardship and environmental outputs and not as a market regulation instrument because for that the stewardship instruments are too expensive and are barely effective. Neither should these policies be used only as income transfer instruments, without delivering benefits to society.

If agri-environmental policy is to expand further as part of agricultural policy, as we accept that it should, the recommendations from this work, based on new empirical and methodological findings, may assist that process. It will contribute if it succeeds in persuading policy-makers of the successes and failures of recent policies and the options as yet untried. There can be no

Box 9.1. Policy recommendations with respect to future application of CSPs

- CSPs should be more closely targeted to specific environmental problems, paying only for real environmental outcomes and not simply because farmers are respecting the law;
- There is scope for application of the beneficiary-pays principle in a number of cases where the produced EGSs are (potentially) marketable. The role of policy should in this case be limited to designation of property rights and temporary incentives to get farmers started to develop the market;
- Some useful indicators have been developed showing the adherence of applied CSPs with general policy principles. The analysis distinguishes seven types of policy and indicates that for a number of them adherence can be doubted or at least be questioned (in particular the PPP). The sustainability of practices after the end of contracts is certainly a problem. Efforts are necessary to make benefits more persistent;
- There are clearly trade-offs to be made between the environmental benefits, the compensations to farmers and administration costs;
- Transaction costs are an essential part of policy application. The analysis reveals important differences in transaction costs between similar policies and thus possible scope to economise on them;
- Greater transparency with regard to administrative costs is required as a safeguard against inappropriate public policy spending and losses of social welfare;
- It is relatively easy to transfer payments to landholders, but much more difficult to ensure that environmental management conditions are followed in return. Countryside stewardship schemes tend to be more complex, frequently involving farmer-specific negotiations on participation and substantial professional input from project officers. But some schemes are more rigorous in their conservation management requirements than others. Policies should take all costs into account and evaluate whether social benefits (extra production of EGS or extra prevention of pollution above existing legal obligations) are higher than total costs (compensation + transaction costs);
- Private transaction costs should also be taken into account when designing policies. The farm survey reveals that farmers perceive extra administration, labour and investments when applying CSPs on their farm. This might be a reason for lower uptake of certain policies;
- It must be stressed that exhaustive scheme evaluation is still premature, given the short period such programmes have been implemented but when evaluated the underlying goals of policies, environmental improvements, are the appropriate focus;
- Information seems to be important to increase the stewardship role of farmers;
- The farm survey indicates that in certain cases farmers are paid for conservation efforts they otherwise would make without compensation. However, the payments may be needed to ensure attainment of longer term objectives;
- Part of the compensation paid may be used for intensification of other parts of the farm, offsetting both output reduction and environmental conservation efforts. A possible solution is globalising payments at farm level and applying the cross compliance option of Agenda 2000;
- More coherence between agri-environmental policies and general CAP regimes (full decoupling) is desirable in order to avoid using CSPs merely to counteract the negative externalities of CAP policies;
- There is a need for evaluation procedures to estimate the economic benefits of agri-environmental payments as well as to check conformity with WTO rules.

doubt that much further work will be needed if the new policy opportunities offered by Agenda 2000 and beyond, further expanding the possibilities of Article 19 of Regulation 797/85 and Regulation 2078/92, are to play a full role in restoring an appropriate level of countryside stewardship. Whatever policies are introduced, their designers must recognise the potential impact of these new policies on existing markets and *vice versa* if they are to succeed in the long term.

References

International Association of Water Studies (1996) *Assessing the benefits of surface water quality improvements: a manual.* London.
Bromley, D. W. (1991) *Environment and Economy: property rights and public policy.* Basil Blackwell, Oxford.
EC (1990) *Draft Council Decision on the setting up of a Model Scheme for Information on Rural Development Initiatives and Agricultural Markets.* Document 7320/90 (COM (90) 230).
Hodge, I. (1991) The provision of public goods in the countryside: How should it be arranged? In: N. Hanley (Ed.), *Farming and the countryside.* CAB International, Wallingford.
Hofreither, M. F. (1998) *Sustainability in agriculture—Tensions between ecology, economics and social sciences.* Paper presented at the International Konferenz der IER und der Universität Hohenheim, Stuttgart.
Latacz-Lohmann, U. (1998) Mechanisms for the provision of public goods in the countryside. In: S. Dabbert, A. Dubgaard, L. Slangen and M. Whitby (Eds.), *The Economics of Landscape and Wildlife Conservation.* CAB International, Wallingford.
Weaver, R. D. (1998) Private provision of public environmental goods: policy mechanisms for agriculture. In: S. Dabbert, A. Dubgaard, L. Slangen and M. Whitby (Eds.), *The Economics of Landscape and Wildlife Conservation.* CAB International, Wallingford.
Williamson, O. E. (1985) *The economic institutions of capitalism: firms, markets, relational contracting.* Free Press, New York.

Annexe: Overview of inventoried policies

Legend for reading the Table

In the next table all policies inventoried (closing date of inventory, 31 December 1996) are listed and following characteristics indicated:

- Number of the policy in the inventory
- Policy name in English and original language
- Year of issue
- Number of farms enrolled in the last available year (in most cases 1996)
- Area in hectares under the policy in the last available year (in most cases 1996)
- Funds assigned in EURO (total over all available years since year of issue)
- Chap. 2a = Primary objectives of the policies (cf. Table 2.3)
    1. Agricultural crop and rural landscape conservation
    2. Natural environment conservation
    3. Wildlife and biodiversity conservation
    4. Soil conservation
    5. Recreation
    6. Reduction of negative impacts of agriculture
    7. Support to local production and quality labels
    8. Afforestation of agricultural land
    – identification of main objective not possible
- Ch. 2b = Technical aspects of the intended EGS (location on PPC – see Table 2.1 and Fig. 2.2.)
    1. By-product with social costs (ERBDs)
    2. Production factor
    3. Incidental by-product
    4. Intentional product
    5. Main product
    6. Only product
    7. Answer not possible
- Ch. 3 = Classification according to the groups identified by the typology in Chapter 3 (Fig. 3.5.) in bold: CSPs situated in the kernel of their group
- Ch. 4 = CSPs analysed in Chapter 4 are identified by an X (measures grouped under one policy are indicated under the policy name in bold)
- Ch. 7 = Classification of CSPs according to the instruments used (see section 7.2)
    1. Ban of inputs (mineral fertilizer, herbicides, pesticides, water)
    2. Input restriction (mineral fertilizer, herbicides, pesticides or water)
    3. Reduction/limitation of livestock and/or reduction of farm wastes
    4. Upkeep of agricultural land
- Ch. 8 = Classification of CSPs according to the clusters identified in Chapter 8 (Table 8.3)
    1. Cluster 1
    2. Cluster 2
    3. Cluster 3
    4. Cluster 4
- n.a. means data not available or not applicable (e.g. in case of demonstration or extension projects)
- no indication in the last 4 columns means that the CSP has not been used or classified

Austria

	Policy	Year of issue	Number of farms	Area (ha)	Funds (Euro)	Ch. 2a	Ch. 2b	Ch. 3	Ch. 4	Ch. 7	Ch. 8
801	ÖPUL Summary ÖPUL—Übersicht	1995	n.a.	n.a.	n.a.	–	7			1;2;3	
802	Elementary support Elementarförderung	1995	169,955	2,302,968	118,000,000	–	3	**G4**		2;3	2
803	Organic farming Biologische Wirtschaftsweise	1995	15,844	197,952	50,740,000	–	3	**G2**	X	1;3	2
804	Reduction of agro-chemicals, whole farm level Verzicht ertragssteigernde Betriebsmittel (Betrieb)	1995	37,718	309,729	50,020,000	6	1	G4		1	
805	Integrated fruit production Integrierte Produktion Obstbau	1995	3,290	10,156	5,500,000	–	3			2	
806	Integrated wine production Integrierte Produktion Weinbau	1995	15,970	42,520	26,000,000	6	1			2	
807	Integrated production of ornamental plants Integrierte Produktion Zierpflanzen	1995	76	479	180,000	6	1			2;3	
808	Extensive grassland production Extensive Grünlandbewirtschaftung	1995	10,848	111,647	20,330,000	1	3			1;3	3
809	Crop rotation measures Fruchtfolgeförderung	1995	53,987	908,764	98,630,000	–	3	G7			3
810	Extensive bread-grains production Extensiver Getreidebau	1995	78,675	330,202	46,000,000	6	1	G2		2;3	2
811	Reduction of agri-chemicals Verzicht auf best. ertragssteigernde Betriebsmittel	1995	78,675	249,215	24,750,000	6	1	G4		2	2
812	Reduction of agri-chemicals Verzicht auf Handelsdünger und Pflanzenschutzmittel	1995	45,841	246,571	33,810,000	6	1	G3		2;3	2
813	Cutting grassland according to time-schedule Schnittzeitauflagen	1995	2,745	6,349	950,000	–	4	G3			2
814	Erosion control, fruit production Erosionsschutz Obstbau	1995	2,814	6,062	710,000	4	2	G7			
815	Erosion control in vineyards Erosionsschutz Weinbau	1995	2,814	3,315	590,000	4	2	G7			

Annexe: overview of inventoried CSPs   195

ID	Name	Year								
816	Erosion control in arable land Erosionsschutz Ackerbau	1995	273		50,000	4	2	G7		
817	Endangered livestock breeds Seltene Tierrassen	1995	3,329	0	1,660,000	3	4	G7		1
818	Moving of steep plots Mahd von Steilflächen und Bergmähdern	1995	58,310	232,389	46,600,000	1	4		4	2
819	Grazing Alpine grassland Alpungs- u. Behirtungsprämie	1995	8,771	267,591	20,000,000	2	3	G5	1;2;4	2
820	Management of ecologically vulnerable areas Pflege ökologisch wertvoller Flächen	1995	41,699	35,323	10,800,000	2	5	G1	1;2;3;4	2
821	Cultivation of endangered crops Seltene landw. Kulturpflanzen	1995	11	18	5,800	3	4			
822	Management of abandoned forests Pflege aufgegebener forstwirtsch. Flächen	1995	164	665	210,000	2	4		4	
823	Landscape elements on set aside land 20jährige Stillegung K1	1995	164	665	80,000	1	6		1;3	
824	Ecological objectives 1 Ökolog. Ziele a. konjunkt. Stillegungsflächen K3	1995	763	536	230,000	6	1	G1	1;3	
825	Ecological objectives 2 Ökolog. Ziele a. konjunkt. Stillegungsflächen K2	1995	1,874	4,595	420,000	6	1	G1	1;3	
826	Extension Bildungsmaßnahmen	1995	n.a.	n.a.	n.a.	–	7			
827	Water Act Wasserschutzgesetz	1956	n.a.	n.a.	n.a.	6	3		2;3	
828	Livestock Act Viehwirtschaftsgesetz	1994	n.a.	n.a.	n.a.	6	4		3	
829	Fertiliser Act Düngemittelgesetz	1995	n.a.	n.a.	n.a.	6	4			
830	Soil Conservation Act Bodenschutzgesetz Oberösterreich	1991	n.a.	n.a.	n.a.	4	4		2;3	
831	Soil Protection Act Bodenschutzgesetz Niederösterreich	1991	n.a.	n.a.	n.a.	4	4		2;3	

Austria continued

	Policy	Year of issue	Number of farms	Area (ha)	Funds (Euro)	Ch. 2a	Ch. 2b	Ch. 3	Ch. 4	Ch. 7	Ch. 8
832	Soil Protection Act Bodenschutzgesetz Burgenland	1990	n.a.	n.a.	n.a.	4	4			2;3	
833	Soil conservation in Steiemark Bodenschutzgesetz Steiermark	1987	n.a.	n.a.	n.a.	4	4			2;3	
834	Ecological Minimum Standards VO zu den ökologischen Mindestkriterien	1995	n.a.	n.a.	n.a.	–	3			2;3	4
835	Land consolidation Flurverfassungsgesetze	1951	n.a.	92	1,850,000	1	2				
836	Regional Project Steimark Regionalprojekt Steiermark F1	1996	n.a.	n.a.	n.a.	6	3	G4		2;3	
837	Regional project Steimark Regionalprojekt Steiermark 22	1996	n.a.	n.a.	n.a.	6	3			2;3	
838	Eco-points Niederosterreich Ökopunkteprogramm Niederösterreich	1995	582	9,921	3,400,000	–	3	G4	X	2;3	
839	Promotion of extensive meadows Wiesenprogramm Burgenland	1995	n.a.	2,063	138,000	3	4			1;2;3	
840	Promotion of extensive fruit production Streuobstwiesenprogramm Burgenland	1995	n.a.	901	133,000	1	4			1;2;3	
841	Reduction of livestock density Reduktion des Viehbestandes	1995	0	0	0	6	3			3	

Belgium

	Policy	Year of issue	Number of farms	Area (ha)	Funds (Euro)	Ch. 2a	Ch. 2b	Ch. 3	Ch. 4	Ch. 7	Ch. 8
101	2078—Aid to Organic Farming Steunregeling biologische teeltmethoden	1994	132	3,444	1,385,465	6	1	G2	x	1	1
102	2078—Demonstration projects—monitoring and warning services Demonstratieprojecten : waarnemings— en waarschuwingsdiensten	1996	4	n.a.	294,002	6	3			2	
103	2078—Demonstration projects—experimental fields Demonstratieprojecten : verminderd gebruik van meststoffen en gewasbeschermingsmiddelen	1996	14	n.a.	285,078	6	3			2	
104	2078—Demonstration projects—organic farming Demonstratieprojecten : biologische landbouw	1996	2	n.a.	81,805	6	4			1	
105	Premium for afforestation on agricultural land Subsidies voor de bebossing van landbouwgrond	1996	4	12	n.a.	8	5				
106	Objective 5a of the structural policies: method 1, aid to farm improvement plan Steun aan investeringen ter verbetering van de structuur van het bedrijf (bedrijfsverbeteringplan)	1992	139	0	1,903,860	6	3			2;3	
107	Objective 5a of the structural policies: method 6, Aid to investments aimed at protecting/improving the environment Regionale steun voor investeringen gericht op de bescherming en verbetering van het leefmilieu en het dierenwelzijn	1992	113	0	3,847,989	6	3			2;3	
108	Landscape Decree Decreet houdende de bescherming van landschappen	1996	0	36,000	n.a.	3	5			2;3	
109	Vegetation Decree Instelling van een vergunningsplicht voor de wijziging van vegetatie en van lijn- en puntvormige elementen	1992	n.a.	n.a.	0	1	4				
110	Manure Decree—implementation of Nitrate Directive 91/676 Decreet inzake de bescherming van het leefmilieu tegen de verontreiniging door meststoffen (Mestdecreet)	1994	44,467	633,896	0	6	1				

Belgium continued

	Policy	Year of issue	Number of farms	Area (ha)	Funds (Euro)	Ch. 2a	Ch. 2b	Ch. 3	Ch. 4	Ch. 7	Ch. 8
111	Economic compensations in the framework of the 'Manure Decree' Economische vergoedingen in het kader van het Decreet inzake de bescherming van het leefmilieu tegen de verontreiniging door meststoffen	1996	10,000	130,000	24,789,352	2	2				
112	Land consolidation with concern for landscape preservation Ruilverkaveling 'Nieuwe Stijl'	1996	4,000	26,000	8,552,327	1	2	G4	x	3	
113	Support agreements in the province of Limburg Ondersteuningsovereenkomsten in de provincie Limburg	1992	56	n.a.	70,783	3	2				
114	Plantation in yard in the Province of East-Flanders Erfbeplantingsactie in de provincie Oost-Vlaanderen	1992	79	n.a.	32,322	1	4			2;3	
115	Small-scale water purification on farms in the Province of East-Flanders Replantation and maintenence of pillard-willows and ponds in the Province of East-Flanders Kleinschalige afvalwaterzuivering op landbouwbedrijven in de Provincie Oost-Vlaanderen	1996	4	n.a.	7,437	6	7			2;3	2
116	Plantation and maintenence of pillard-willows and ponds in the Province of East-Flandres Provinciaal reglement inzake onderhoud en (her)aanleg van knotwilgen en veedrinkpoelen in Oost-Vlaanderen	1994	14	n.a.	17,302	2	3	G7	X		
117	Plantation in yard in the Regional Landscape (Natural Park) Noord-Hageland Erfbeplanting in het Regionaal Landschap Noord-Hageland	1996	40	n.a.	25,310	1	6				
118	Ponds and waterside landscapes in the Regional landscape (natural Park) Noord-Hageland (Her)aanleg van poelen en waterkanten in het Regionaal Landschap Noord-Hageland	1996	22	n.a.	36,044	2	6				

Annexe: overview of inventoried CSPs

No	Title	Year							
119	Maintenance of small landscape elements (hedges, tree rows, wooded banks, pollards) at the municipality Zoutleeuw / Onderhoud van kleine landschapselementen in de Gemeente Zoutleeuw	1995	20	n.a.	9,915	1	6		
120	Research on the erosion control in the Regional Landscape West-Vlaamse heuvels / Het onderzoeken van de erosieproblematiek in de West-Vlaamse Heuvels en het opzetten van proefprojecten bij landbouwers	1996	3	n.a.	12,395	4	2		
121	Plantation action in the Province of West-Flanders Beplantingsactie 'De Groene Zetel' in West-Vlaanderen	1995	336	n.a.	36,027	1	6		
122	Reduction and recycling of waste water on dairy farms in the Province of West-Flanders / Reductie en hergebruik van afvalwater op melkveehouderijen in de provincie West-Vlaanderen	1994	7	n.a.	2,520	6	2	3	
123	Pond project in the Region 'Zwinstreek' / Poelenproject 'Zwinstreek'	1996	15	n.a.	3,718	1	2		
124	Tax exemption for agricultural used water considered as domestic used water / Exemption de la taxe sur le déversement des eaux usées agricoles assimilées aux eaux usées domestiques	1992	7,499	273,366	5,690,547	6	1	G4	1
125	Management of nature reserves by RNOB / Contrats d'entreprise	1996	50	200	2,287	2	5	G3	4
126	Management agreements with farmers by RNOB / Conventions de jouissance limitée à titre gratuit	1996	0	0	0	3	5	G3	1
127	Municipality Nature Development Plan / PCDN 'Plan communal de Développement de la Nature'	1994	n.a.	n.a.	312,353	3	4	G3	
128	Land consolidation: plots regrouping / Remembrement regroupement des parcelles	1996	2,026	14,666	7,932,593	–	2	G7	
129	Land consolidation: sites evaluation and management / Remembrement plan d'évaluation et d'aménagement des sites	1996	1,173	8,699	91,721	–	4	G5	

Belgium continued

	Policy	Year of issue	Number of farms	Area (ha)	Funds (Euro)	Ch. 2a	Ch. 2b	Ch. 3	Ch. 4	Ch. 7	Ch. 8
130	Quality labels for 'blanc-bleu' beef, chicken and pigs Labels 'Blanc-Bleu fermier', 'Poulet de chair' et 'Porc fermier'	1996	932	n.a.	247,894	7	3	**G2**	X		
	Agri-environmental programme in the Walloon region—Regulation 2078/92 (measures 131–145)								X		
131	2078—preservation and maintenance of hedges and woodland strips MAE—Maintien et entretien des haies et bandes boisées	1995	784	7,210	188,155	1	4	G7		4	2
132	2078—late mowing Fauche tardive et diversification des semis en prairie temporaire	1995	142	1,022	142,542	3	4	G3		1;2;3	3
133	2078—grass field margin: replacement of cropland by a grass field margin MAE—tournières de conservation	1995	12	42	10,412	6	1	G3		1;3	2
134	2078—extensive field margin MAE—tournières de conservation	1995	2	10	1,289	6	1			1;2;3	
135	2078—grass field margin – replacement MAE—tournières de conservation	1995	1	2	595	6	1			1;3	
136	2078—reduction of livestock grazing density MAE—Maintien de faibles charges en bétail	1995	50	1,864	101,639	6	1	G7		3	4
137	2078—reduction of inputs in cereals 1 MAE—réduction d'intrants en céréales	1995	4	45	3,718	6	1			1;3	
138	2078—reduction of inputs in cereals 2 MAE—réduction d'intrants en céréales	1995	3	61	5,454	6	1			2	
139	2078—reduction of herbicide in maize 1 MAE—réduction et localisation des herbicides en maïs avec mécanisation du désherbage	1995	1	7	1,011	6	1			2	
140	2078—reduction of herbicide in maize 2 MAE—sous-semis en maïs	1995	1	5	744	6	1				

Annexe: overview of inventoried CSPs

#	Name	Year	C1	C2	C3	C4	C5	C6	C7	C8
141	2078—winter green cover / MAE couverture du sol avant culture de printemps	1995	2	85		6	1			
142	2078—very late mowing / MAE—fauche très tardive avec limitation des intrants	1995	5	34	8,428	3	5		1;3	3
143	2078—wetland conservation / Mesures conservatoires en zones humides	1995	5	26	1,289	2	5		1;3	3
144	2078—conservation of farms 1 / MAE—Fermes de conservation	1995	1	1	62	7	2		2	
145	2078—conservation of farms 2 / MAE—Fermes de conservation	1995	1	1	124	7	2		2	
146	Objective 5b: plantation and maintenance of hedges and lines of trees / Objectif 5b: plantation et entretien des bocages	1995	20	n.a.	31,226	1	4	G5	4	
147	Objective 5b: marketing of hedges and line of trees maintenance work / Objectif 5b: atelier environnement	1995	120	n.a.	177,793	1	4		4	
148	2080—conversion of agricultural land into forest / Octroi d'une subvention aux propriétaires particuliers pour la régénération—Boisement des terres agricoles—Règl. 2080/92	1996	9	35	2,169	8	5	G7		4
149	2080—incentive scheme for hedge and wind break plantation / Octroi d'une subvention à la plantation de haies - Règl. 2080/92	1996	1	2	331	6	1	X	1	
150	2080—Compensations for afforested agricultural land / Prime destinée à compenser la perte de revenus découlant du boisement des terres agricoles en application du règl. (CEE) no 2080/92	1996	4	12	n.a.	8	5			
151	Beef extension premium / Prime d'extensification bovine—Règl.(EEC) No 2066/92	1993	6,400	96,200	14,417,809	6	1		3	2
152	Objective 5b: promotion and development and structural adjustment of rural areas / PDZR—valorisation des productions animales	1996	101	7,160	536,813	7	3		2	2

Belgium continued

	Policy	Year of issue	Number of farms	Area (ha)	Funds (Euro)	Ch. 2a	Ch. 2b	Ch. 3	Ch. 4	Ch. 7	Ch. 8
153	Contrat de rivière—ecologically sound action programme on a river basin (Semois) Contrat de rivière (Semois)	1993	n.a.	n.a.	195,840	2	7			3;4	
154	Agricultural investment funds: regime in favour of the preservation and improvement of the environment FIAW regime national en faveur de la protection et de l'amelioration de l'environnement	1994	n.a.	n.a.	n.a.	6	2			3	
155	Integrated method for pip fruit production Méthode de production intégrée pour fruits à pépins (pommes et poires)	1988	35	750	189,862	6	1	**G2**		2	

Annexe: overview of inventoried CSPs 203

France

	Policy	Year of issue	Number of farms	Area (ha)	Funds (Euro)	Ch. 2a	Ch. 2b	Ch. 3	Ch. 4	Ch. 7	Ch. 8
701	Grassland premium *Prime à l'herbe*	1993	99,691	5,393,500	748,164,677	3	2	**G6**	X	3;4	1
702	Taking land out of agricultural production Retrait à long terme	1993	119	474	1,403,526	6	1	G3	X	1;2;3	1
703	Conversion from arable to extensive grassland Reconversion des terres arables	1993	3,254	17,118	35,824,857	4	4	G6	X	1;2;3	1
704	Rearing of threatened breeds Protection des races menacées	1993	1,972	n.a.	4,921,008	3	2	G7	X		4
705	Reduced livestock densities Diminution du chargement du cheptel	1993	1,292	n.a.	34,751,515	6	1	G7	X	3	2
706	Conversion to organic farming Reconversion en agriculture biologique	1993	2,600	66,290	57,465,620	7	3	**G2**	X	1;2;3	3
707	Reduced use of agri-inputs Réduction des intrants	1993	2,596	60,578	60,675,395	6	1	**G4**	X	1;2;3;4	2
708	Local programmes and other regional priorities Opérations locales—OL	1993	32,874	649,173	301,154,177	1	4		X	4	
709	Local programme planting hedgerows OL bocage à ormes	1994	75	1,031	628,822	3	2	G5		1;2;3	1
710	Local programme wetlands OL Marais du Cotentin	1991	300	3,947	278,236	7	3	**G1**		1;2;3	1
711	Local programme uplands OL Monts d'Arrée	1993	170	2,247	834,830	6	1	**G1**		2;3	1
712	Local programme Normandie-Maine OL Normandie-Maine	1993	315	4,600	291,806	6	3	**G1**		2;3	3
713	Sustainable development scheme Plans de développement durable	1991	800	n.a.	n.a.	6	1	**G6**		2;3	
714	Scheme for improving fertilisation FERTI-MIEUX actions Fertimieux	1990	25,000	1,500,000	n.a.	1	4	G4		4	

France continued

	Policy	Year of issue	Number of farms	Area (ha)	Funds (Euro)	Ch. 2a	Ch. 2b	Ch. 3	Ch. 4	Ch. 7	Ch. 8
715	European initiative for integrated farming réseau FARRE—Forum pour une Agriculture Raisonnée et Respectueuse de l'Environnement	1993	57	6,500	0	8	4	**G4**		2;3;4	
716	Agricultural pollution monitoring scheme Programme maîtrise des pollutions d'origine agricole	1993	16,196	n.a.	452,000,000	6	1	G4		2	3
717	Countryside maintenance funding sheme Fonds de gestion de l'espace rural	1995	n.a.	n.a.	187,600,000	3	2	**G5**		1;3	1
718	Afforestation Boisement des terres agricoles	1992	540	5,550	980,308	8	2	G4		2	1
719	Wine label AOC Bordeaux AOC Vins de Bordeaux	1936	12,500	112,200	0	7	2	**G2**		2;3;4	
720	Cheese label AOC Beaufort AOC Fromage de Beaufort	1968	800	450,00	0	7	2	**G2**		1;3	
721	Coast and lake shore protection Conservatoire du Littoral	1986	0	2,067	n.a.	2	4	**G6**		1;3;4	
722	Set aside for fauna protection Jachère Faune sauvage	1992	4,101	22,124	n.a.	3	5	G1		1;2	3

Germany

	Policy	Year of issue	Number of farms	Area (ha)	Funds (Euro)	Ch. 2a	Ch. 2b	Ch. 3	Ch. 4	Ch. 7	Ch. 8
	MEKA (Reg. 2078/92) Baden-Württemberg (2001–2031)								X		
2001	Extensive use of grassland 1 (< 1.2 LU per ha) Grünlandgrundförderung—Besatzdichtestufen 1	1992	10,161	105,073	32,771,598	6	1	G3		1;3	2
2002	Extensive use of grassland 2(1.2–1.8 LU per ha) Grünlandgrundförderung—Besatzdichtestufen 2	1992	5,628	79,077	16,807,824	6	1			1;3	2
2003	Extensive use of grassland 3 (> 1.8 LU per ha) Grünlandgrundförderung—Besatzdichtestufen 3	1992	5,195	50,750	7,867,864	6	1			1;3	
2004	Grassland with slope gradient 25–50% Förderung der Bewirtschaftung von Steillagen-grünland—25–50%	1992	14,719	57,532	15,168,963	4	3	G5			4
2005	Grassland with slope gradient over 50% Förderung der Bewirtschaftung von Steillagen-grünland—>50%	1992	2,593	5,749	2,861,274	4	3				
2006	Extensive use of grassland: max 2 cuts Extensives Grünland: max 2 Nutzungen	1992	22,427	119,303	6,449,234	1	7	G7			4
2007	Extensive use of grassland: one cut Extensives Grünland: einschürig	1992	5,018	11,790	1,327,379	1	7				
2008	Extensive use of grassland: humid and wet locations Grünlanf (feucht/naß)	1992	7,504	20,465	4,635,409	2	5				
2009	Extensive use of permanent grassland Extensive Grünlandnutzung	1994	5,695	79,678	15,747,407	1	4			1;2;3	
2010	Preservation of meadows with scattered orchard trees Förderung von Streuobstwiesen	1992	39,996	65,618	32,082,835	1	4	G5			4
2011	Preservation of vine growing at steep slopes Förderung Steillagenweinbau	1992	699	300	162,176	1	3				
2012	To rear animals of local breeds in danger of extinction: Vordwälder Rind (Cattle) Haltung und Aufzucht gefährdeter Nutztierrassen	1992	3,749	24,058	5,817,437	3	4	G7			4

Germany continued

	Policy	Year of issue	Number of farms	Area (ha)	Funds (Euro)	Ch. 2a	Ch. 2b	Ch. 3	Ch. 4	Ch. 7	Ch. 8
2013	To rear animals of local breeds in danger of extinction: Hinterwalder Rind (Cattle) Haltung und Aufzucht gefährdeter Nutztierrassen	1992	655	2,949	1,308,985	3	4				
2014	To rear animals of local breeds in danger of extinction: Limpurger Rind (Cattle) Haltung und Aufzucht gefährdeter Nutztierrassen	1992	93	383	106,154	3	4				
2015	To rear animals of local breeds in danger of extinction: Schwarzwalder Fuchs (Horse) Haltung und Aufzucht gefährdeter Nutztierrassen	1992	356	612	255,832	3	5				
2016	To rear animals of local breeds in danger of extinction: Suddeutsches Kaltblut (Horse) Haltung und Aufzucht gefährdeter Nutztierrassen	1992	148	246	132,219	3	5				
2017	To rear animals of local breeds in danger of extinction: Altwuttemberger Pferd (Horse) Haltung und Aufzucht gefährdeter Nutztierrassen	1992	158	284	115,957	3	5				
2018	Renunciation of pesticides and mineral fertiliser in arable farming Verzicht auf Mineral-dünger und chem. Pflanzenschutz im Ackerbau	1992	5,659	60,052	19,022,343	6	1		1		
2019	Renunciation of growth regulators in wheat cultivation Verzicht auf Wachstumsregulatoren	1992	27,403	116,858	44,705,586	6	1	G4			2
2020	Renunciation of growth regulators in rye/triticale cutivation Verzicht auf Wachstumsregulatoren	1992	5,105	11,830	3,029,608	6	1				
2021	Enlargement of seed-rows distance to min 17 cm in cereal cultivation Erweiterung des Drillreihenabstands im Getreidebau	1992	12,058	184,356	31,102,915	6	1	G7			2
2022	Biotopes at humid or wet locations Feuchtbiotope	1992	459	436	292,643	2	5			4	
2023	Biotopes at dry locations Trockenbiotope	1992	255	3,027	1,127,259	2	5			4	
2024	Mulch-seeding in arable farming Mulchsaat	1992	6,568	90,420	19,174,785	4	7	G7			4

Annexe: overview of inventoried CSPs

ID	Name	Year									
2025	Special biotopes Sonderbiotope	1992	630	285	734,471	2	5			4	4
2026	Renunciation of herbicides in arable farming Verzicht auf Herbizide im Ackerbau	1992	10,116	29,378	6,737,251	6	1	G7		1	4
2027	Underseeds and catch crops in arable farming and permanent crops Untersaaten und Zwischenfruchtanbau	1992	40,401	278,416	80,692,871	4	7	G7			4
2028	Promotion of organic farming; permanent crops: introduction Förderung Ökolandbau Dauerkulturen; Einführung	1994	92	385	689,067	6	1			1;3	4
2029	Promotion of organic farming; permanent crops: maintenance Förderung Ökolandbau Dauerkulturen; Beibehaltung	1994	162	527	390,330	6	1			1;3	
2030	Promotion of organic farming; grassland and arable land: introduction Förderung Ökolandbau Acker- und Grünland; Einführung	1994	448	12,873	3,816,357	6	1			1;3	
2031	Promotion of organic farming; grassland and arable land: maintenance Förderung Ökolandbau Acker- und Grünland; Beibehaltung	1992	701	19,902	3,044,159	6	1	**G2**		1;3	
	FUL (Reg. 2078/92) Rheinland-Pfalz (2032–2038)								X		
2032	Long-term set aside of arable land: 20 years 20-jährige Ackerflächen-stillegung für ökologische Zwecke	1993	72	236	94,332	2	1	**G3**			4
2033	Integrated farm management in arable farming and permanent cropping Integriert-kontrollierte Wirtschaftsweise im Acker-, Obst-, und Weinbau	1993	818	21,885	4,974,241	6	4				
2034	Promotion of organic farming Förderung Ökolandbau	1993	108	2,381	648,542	6	1				

Germany continued

	Policy	Year of issue	Number of farms	Area (ha)	Funds (Euro)	Ch. 2a	Ch. 2b	Ch. 3	Ch. 4	Ch. 7	Ch. 8
2035	Field margins of arable land Ackerrandstreifenprogramm	1993	11	18	12,053	3	4				
2036	Extensification of grassland: variant 1 Grünlandextensivierung Gesamtbetrieb	1993	966	32,860	4,651,060	6	1	G3			2
2037	Extensification of grassland: variant 2 Grünlandextensivierung Einzelflächen	1993	784	2,998	681,013	6	1	**G3**			4
2038	Preservation of meadows with scattered orchard trees Förderung Streuobstwiesen	1993	385	395	100,477	1	4				
	UL (Reg. 2078/92) Sachsen (2039–2070)										
2039	Integrated farm management Intergriert-kontrollierte Wirtschaftsweise	1994	1,319	427,030	35,668,294	6	1	G4		2;3	2
2040	Additional support I: measures to relief the environment Zusatzförderung I: umweltentlastende Maßnahmen	1994	563	117,029	15,436,628	6	1	**G4**		2;3	4
2041	Additional support II: growing of catch crops Zusatzförderung II: Zwischenfruchtanbau	1994	196	8,765	582,592	4	7	G7			
2042	Additional support II: underseeds Zusatzförderung II: Untersaaten	1994	114	4,006	379,051	4	7				
2043	Additional support II: mulch-seeding Zusatzförderung II: Mulchsaat	1994	64	2,775	112,817	4	3				4
2044	Additional support II: land planting on set aside plots Zusatzförderung II: Begrünung von Stillegungsflächen	1994	197	7,231	480,630	4	2				
2045	Organic farming: arable farming Förderung Ökolandbau Ackerbau	1994	20	567	213,976	6	1			1;2;3	
2046	Integrated management in cultivation of field vegetables Integriert-kontrollierte Wirtschaftsweise im Gemüsebau	1994	60	2,411	1,126,887	6	1	G4		2;3	2

ID	Name	Year								
2047	Integrated management in fruit growing / Integriert-kontrollierte Wirtschaftsweise im Obstbau	1994	48	4,048	3,754,928	6	1		2;3	2
2048	Integrated management in vine growing / Integriert-kontrollierte Wirtschaftsweise im Weinbau	1994	16	54	40,494	6	1		2;3	
2049	Integrated management in hop growing / Integriert-kontrollierte Wirtschaftsweise im Hopfenbau	1994	12	466	230,593	6	1	**G4**	2;3	
2050	Organic farming: vegetables / Förderung Ökolandbau Gemüsebau	1994	5	3	1,585	6	1		1;3	
2051	Organic farming: fruit culture / Förderung Ökolandbau Obstbau	1994	5	12	15,032	6	1		1;3	
2052	Organic farming: viniculture / Förderung Ökolandbau Weinbau	1994	1	3	4,295	6	1		1;3	
2053	To rear animals of local breeds in danger of extinction: Vogtlandisches Rotrich (Cattle) / Haltung und Aufzucht gefährdeter Nutztierrassen	1994	11	63	8,181	3	4			
2054	To rear animals of local breeds in danger of extinction: Kaltblut (Horse) / Haltung und Aufzucht gefährdeter Nutztierrassen	1994	217	350	72,092	3	4			
2055	To rear animals of local breeds in danger of extinction: sheep, goats / Haltung und Aufzucht gefährdeter Nutztierrassen : Schafe, Ziege	1994	20	281	6,948	3	4			
2056	Conversion of temporary grassland into permanent grassland / Beibehaltung der Grünlandnutzung auf Wechselgrünland	1994	305	12,729	994,115	6	1		3	
2057	Grassland management with reduced input / Grünlandnutzung mit reduziertem Produktions-mitteleinsatz	1994	815	42,986	3,224,858	6	1		2;3	
2058	Late mowing of grassland / Späte Schnittnutzung	1994	578	8,335	2,353,579	6	1	**G3**	1;3	

Germany continued

	Policy	Year of issue	Number of farms	Area (ha)	Funds (Euro)	Ch. 2a	Ch. 2b	Ch. 3	Ch. 4	Ch. 7	Ch. 8
2059	Conversion of arable land into extensive grassland Umwandlung von Ackerland in Extensivgrünland	1994	163	1,341	663,555	6	1	**G3**		1;3	4
2060	Long term set-aside for development of biotopes 20-jährige Flächenstillegung zum Zwecke der Biotopentwicklung	1994	19	98	60,332	3	5	**G3**		1;3	
2061	Upkeep of abandoned farmland Pflege aufgegebener landwirtschaftlicher Flächen	1994	31	524	163,491	1	5			1;3;4	
2062	Board-pasturing (cattle and sheep) Förderung der Pensionsweidehaltung von Rindern und Schafen	1994	18	1,326	239,658	1	2			3	
2063	Late mowing of grassland (> 30.6) Späte Schnittnutzung	1994	494	8,126	3,226,456	6	1			1;3	4
2064	Field margins programme Ackerrandstreifenprogramm	1994	47	424	370,584	3	4	**G3**		1;3	
2065	Maintenance of meadows with scattered orchard trees Pflege von Streuobstwiesen	1994	3,268	2,037	820,112	1	4			3	
2066	Maintenance of ponds Teichpflege	1994	123	7,507	4,248,268	2	2			1;3	
2067	Extensive management of grassland Extensive Grünlandnutzung	1994	138	2,084	671,567	6	1	**G3**		1;3	4
2068	Single liming for soil melioration Einmalige meliorative Kalkung	1994	9	308	79,508	4	2			3	
2069	Conversion of arable land into permanent grassland Umwandlung von Ackerland in Dauergrünland	1994	46	345	342,054	6	1			1;3	
2070	Planting of scattered orchard trees Anlegen und erneuern von Streuobstwiesen	1994	15	876 trees	25,118	1	4			3	
	KULAP (Reg. 2078/92) Brandenburg (2071–2084)										
2071	Organic farming in permanent cropping Förderung Ökolandbau: Dauerkulturen	1994	n.a.	515	587,708	6	1			1;3	

Annexe: overview of inventoried CSPs

2072	Organic farming: arable farming/grassland Förderung Ökolandbau: Acker- und Grünland	1994	n.a.	6,512	1,657,650	6	1	1;3	
2073	Grassland at humid locations Feuchtwiesenprogramm	1994	n.a.	1,628	652,306	2	5	G1	2;3
2074	Grassland at wet locations Feuchtwiesenprogramm(Überflutungsgrünland)	1994	n.a.	11,534	4,914,487	2	5		1;3
2075	Upkeep of abandoned grassland Pflege von brachliegendem Grünland	1994	n.a.	4,620	1,873,340	1	5	G1	1;4
2076	Extensive management of grassland Extensive Grünlandnutzung	1994	n.a.	95,499	26,983,256	1	4	**G3**	1;2;3
2077	Conversion of arable land into extensively managed grassland Umwandlung von Ackerland in extensiv genutztes Grünland	1994	n.a.	10,843	5,453,195	6	1		1;2;3
2078	Renunciation of pesticides and mineral fertiliser in arable farming Verzicht auf Mineraldünger und chem. Pflanzenschutz im Ackerbau	1994	n.a.	15,544	3,351,284	6	1		1;3
2079	Renunciation of mineral fertiliser in arable farming Verzicht auf Mineraldünger im Ackerbau	1994	n.a.	6,570	792,959	6	1		1;3
2080	Renunciation of herbicides in arable farming Verzicht auf Herbizide im Ackerbau	1994	n.a.	9,756	1,457,570	6	1	G7	1;3
2081	Renunciation of pesticides and mineral fertiliser in permanent cropping Verzicht auf Mineraldünger und chem. Pflanzenschutz in Dauerkulturen	1994	n.a.	342	412,471	6	1		1;3
2082	Renunciation of mineral fertiliser in permanent cropping Verzicht auf Mineraldünger in Dauerkulturen	1994	n.a.	2	727	6	1		1;3
2083	Renunciation of chemical herbicides in permanent cropping Verzicht auf Herbizide in Dauerkulturen	1994	n.a.	7	1,709	6	1		1;3
2084	Renunciation of chemical herbicides in orchards Verzicht auf Herbizide im Obstbau	1994	n.a.	27	2,457	6	1		1;3

Germany continued

	Policy	Year of issue	Number of farms	Area (ha)	Funds (Euro)	Ch. 2a	Ch. 2b	Ch. 3	Ch. 4	Ch. 7	Ch. 8
	2078/92—program Niedersachsen (2085–2096)										
2085	Organic farming/grassland: maintenance Förderung Ökolandbau: Grünland; Beibehaltung	1994	79	1,644	242,961	6	1			1;3	
2086	Organic farming/grassland: introduction Förderung Ökolandbau: Grünland; Einführung	1994	n.a.	1,455	1,003,446	6	1			1;3	
2087	Organic farming/arable: introduction Förderung Ökolandbau: Ackerbau; Einführung	1994	n.a.	653	813,927	6	1			1;3	
2088	Organic farming/arable: maintenance Förderung Ökolandbau: Ackerbau, Beibehaltung	1994	94	4,543	557,431	6	1			1;3	
2089	Organic farming/permanent crops: maintenance Förderung Ökolandbau: Dauerkulturen, Beibehaltung	1994	11	124	113,547	6	1			1;3	
2090	Organic farming/permanent crops: introduction Förderung Ökolandbau: Dauerkulturen, Einführung	1994	1	1	196,978	6	1			1;3	
2091	Reduction of herbicides and mineral fertilisers in arable farming: introduction Verzicht auf Mineraldünger und chem. Pflanzenschutz im Ackerbau; Einführung	1994	8	182	115,285	6	1			1;3	
2092	Reduction of herbicides and mineral fertilisers in permanent cropping Verzicht auf Mineraldünger und chem. Pflanzenschutz in Dauerkulturen	1994	6	76	20,686	6	1			1;3	
2093	Long term set aside of arable land: 20 years 20-jährige Ackerflächenstillegung	1995	36	49	31,956	2	6			1;3	
2094	To rear animals of local breeds in danger of extinction Haltung und Aufzucht und Aufzucht gefährdeter Nutztierrassen	1995	31	92	107,883	3	4				
2095	Extensive management of arable land and grassland Extensive Acker- und Grünlandnutzung	1995	35	234	56,242	6	1			1	
2096	Protection of grassland at humid locations Feuchtwiesenschutz-programm	1995	300	1,776	591,565	2	5			1;2;3	

2078/92—Schleswig-Holstein (2097–2110)

ID	Name	Year						G1	1;2;3	
2097	Protection of grassland-biotopes Wiesen- und Weiseökosystemschutz	1993	925	6,988	1,103,426	2	5		1;2;3	4
2098	Protection of humid grasslands Feuchtgrünlandschutz	1993	42	117	109,901	2	5		1	
2099	Protection of dry grassland Schutz von trockenem Magergrünland	1993	1	6	7,056	2	5		1	
2100	Field margins of grassland Grünlandrandstreifen / Ufer	1993	11	17	9,263	3	5			
2101	Set aside of arable land Ackerbrache	1993	123	677	1,096,696	6	1		1	4
2102	Field margins/riverside Ackerrandstreifen / Ufer	1993	24	46	79,577	3	5		1	4
2103	Programme of 'wild herbs' on arable field margins Ackerrandstreifen mit Ackerwildkräutern	1993	3	8	9,265	3	4		1	
2104	Conversion of arable land into extensively managed grassland Umwandlung von Ackerland in Extensivgrünland	1993	46	413	218,354	6	1		1	
2105	Extensification of grassland Grünlandextensivierung	1993	23	471	69,105	6	1		1	
2106	Renunciation of pesticides and mineral fertilisers in arable farming Verzicht auf chem. Pflanzenschutz und Mineraldünger im Ackerbau	1993	n.a.	n.a.	n.a.	6	4		1;3	
2107	Extensification of grassland Grünlandextensivierung	1993	75	3,257	878,410	6	1		1;3	
2108	Organic farming: arable land Förderung Ökolandbau: Ackerbau	1993	40	827	155,527	6	1		1;3	
2109	Organic farming: grassland Förderung Ökolandbau: Grünland	1993	27	374	73,203	6	1		1;3	

Germany continued

	Policy	Year of issue	Number of farms	Area (ha)	Funds (Euro)	Ch. 2a	Ch. 2b	Ch. 3	Ch. 4	Ch. 7	Ch. 8
2110	Extensive management of grassland Extensive Grünlandnutzung	1993	96	3,728	508,485	6	1			1;3	
2111	National Parks Nationalparke	1978	n.a.	726,500	n.a.	2	5				
2112	Nature Reserve Areas Naturschutzgebiete	?	n.a.	684,503	n.a.	2	5				
2113	Nature Parks and Landscape Protection Areas Naturparke und Landschaftsschutzgebiete	1958	0	4,489,900	n.a.	5	3				
2114	Fertilizer Regulation Düngeverordnung	1996	± 500,000	17,200,000	0	6	3				
2115	Regulation on Water Protection Areas and Compensations SchALVO Schutzgebiets- und Ausgleichsverordnung	1988	27,602	305,500	182,187,340	6	1	G4	X		4
2116	Compensation Payments in LFA Ausgleichszulage Benachteiligtes Gebiet	n.a.	246,000	9,400,000	2,274,772,000	7	2				
2117	Environmental Aspects of Land Consolidation Programmes Umweltaspekte von Flurbereinigungen	n.a.	n.a.	n.a.	2,126,872	1	7				
2118	Agri-environmental Programmes by Local Authorities in Baden-Württemberg Exemplarischer Überblick über lokale CSPs	n.a.	n.a.	n.a.	n.a.	1	5				

Greece

	Policy	Year of issue	Number of farms	Area (ha)	Funds (Euro)	Ch. 2a	Ch. 2b	Ch. 3	Ch. 4	Ch. 7	Ch. 8
601	Pilot Program for Biologically produced oil kalergia elias & emporia biologikou eleoladou	1993	6	2	364	–	7	**G2**		1	
602	Specific Program Covering Small Islands of Aegean Sea idika metra gia orismena agrotika proionta pros ofelos ton mikron nision tu Egeou Pelagous	1993	0	0	6,706	–	7	**G2**		1;2	
603	Program for the Support of Forests Activities in Agriculture enishisi ton metron gia ta dasi ston tomea tis georgias	1993	8,700	9,200	29,935	–	7	**G5**			
604	Biological agriculture programma biologikis georgias	1996	873	3,600	1,435	6	1	**G2**	X	1	2
605	Genetic Erosion of Animal Production programma diatirisis fylon paragogikon zoon pu apeiloynte apo genetiki diavrosi	1997	n.a.	0	n.a.	3	2	G2		3	
606	Nitrate reduction Prgramma gia ti miosi tis nitroripansis georgikis proelefsis ston thesaliko kampo	1995	1,150	9,000	3,531	6	1	G2	X	2	3
607	Long term set aside makrohronia pafsi kaliergias georgikon geon	1996	121	30,000	7,715	3	7	**G1**	X	1	2
608	Genetic Erosion Plant Production Programma diatirisis idon 2 pikilion kalliergumenon fiton pu kindinevun apo genetiki diavrosi	1997	n.a.	n.a.	n.a.	3	2	**G6**		2	
609	Protection of Ecosystems of Specific Ecological Significance Programma prostasias ikotopon idieteris simasias & tis agrias hloridas & panidas	1997	n.a.	n.a.	n.a.	–	2	**G6**		2;3	
610	Leader I—Region of Epirus Leader I—Ipiros AE (Ioannina-Thesprotia)	1993	n.a.	n.a.	n.a.	–	2	G4		1;2;3	
611	Leader I—Region of Amvrakikos Leader I—Periohi Amrakikou kolpu	1992	n.a.	n.a.	4,311	–	2	G4		1;2;3	
612	Leader II—Region of Ioannina Leader II—Ioannina-Thesprotia	1996	n.a.	n.a.	n.a.	–	2	G4		1;2;3	
613	Leader II—Region of Etoloakarnania Leader II—Ditikis Etoloakarnanias	1996	n.a.	n.a.	n.a.	–	2	G4		1;2;3	

Greece continued

	Policy	Year of issue	Number of farms	Area (ha)	Funds (Euro)	Ch. 2a	Ch. 2b	Ch. 3	Ch. 4	Ch. 7	Ch. 8
614	Specific Program of Environmental Protection in the Region of Amvrakikos Idiko Programma Prostasias Periballontos Amvrakikou	1996	n.a.	n.a.	772	–	2	G4		2;3	
615	Specific Program of Environmental Protection In the Region of Aheronta Idiko Programma Prostasias Periballontos Aheronta	1996	n.a.	n.a.	n.a.	–	2	G4		2;3	
616	Leader II—Region of South Epirus Leader II—Notias Ipirou	1996	n.a.	n.a.	n.a.	–	2	G4		1;2;3	
617	Leader II—Region of Florina Leader II—Florinas	1996	n.a.	n.a.	n.a.	–	2	G4		1;2;3;4	
618	Leader II—Region of Elikona Leader II—Elikona	1996	n.a.	n.a.	n.a.	–	2	G4		1;2;4	
619	Leader II—Region of Nafpaktos Leader II—Nafpaktou	1996	n.a.	n.a.	n.a.	–	2	G4		1;2	
620	Regional Operational program of Crete Politamiako Epihirisiako Programma Kritis	1994	n.a.	n.a.	n.a.	–	7	G4			

Italy

	Policy	Year of issue	Number of farms	Area (ha)	Funds (Euro)	Ch. 2a	Ch. 2b	Ch. 3	Ch. 4	Ch. 7	Ch. 8
	Regulation 2078/92 schemes in the Veneto region (measures 401–413)								X		
401	A1–A2 Reduction and maintenane of reduction of agricultural inputs A1–A2 Sensibile riduzione dei concimi e dei fitofarmaci + mantenimento	1994	1,265	15,042	9,713,624	6	1	**G2**		1;2	3
402	A3 Organic agriculture A3 Agricoltura biologica	1994	282	2,355	2,070,023	6	1	**G2**		1	3
403	B1 Introduction and maintenance of extensive agricultural production B1 Introduzione o mantenimento delle produzioni vegetali estensive	1994	71	428	98,775	6	1			2	
404	B2 Conversion of arable into extensive grazings B2 Conversione dei seminativi in pascoli estensivi	1994	3	19	n.a.	6	1			1;3	
405	C Reduction of animal stocking rates C Riduzione della densità del patrimonio bovino per unità di superficie foraggera	1994	2	0	22,878	6	5			3	
406	D1 (a) Maintenance and restoration of natural environments and of landscape elements D1 (a) Azioni di conservazione e/o ripristino di spazi naturali e seminaturali e di elementi dell'agroecosistema e del paesaggio agrario	1994	235	163	471,039	1	3	G6		1	
407	D1(b) Introduction of cover crops D1 (b) Introduzione di colture intercalari destinate al sovescio che consentano il mantenimento della copertura vegetale e l'arricchimento in sostanza organica dei suoli	1994	1	21	3,040	6	1	G7		1;3	
408	D1 (c) Introduction of crops for wildlife feeding D1 (c) Introduzione di colture a perdere per l'alimentazione naturale della fauna selvatica	1994	0	0	0	3	6			1	
409	D2 Breeding of species in danger of extinction D2 Allevamento di specie animali locali in pericolo di estinzione	1994	176	0	188,835	3	3				

Italy continued

	Policy	Year of issue	Number of farms	Area (ha)	Funds (Euro)	Ch. 2a	Ch. 2b	Ch. 3	Ch. 4	Ch. 7	Ch. 8
410	E Restoration of abandoned forest land E Cura dei terreni agricoli e forestali abbandonati	1994	8	249	92,421	4	2			1:4	
411	F Twenty-years set aside F Ritiro dei seminativi dalla produzione per 20 anni	1994	7	54	65,920	2	6			1	
412	G Management of land for access and recreation G Gestione di terreni per l'accesso al pubblico e per le attività ricreative	1994	2	4	1,673	5	5				
413	Information, training and demonstration farms Iniziative di formazione, divulgazione e progetti dimostrativi	1994	n.a.	n.a.	n.a.	–	7				
414	Regional Programme for the application of Reg CEE 2080/92 Programma Pluriennale Regionale (94–96) di applicazione del Reg. CEE 2080/92	1993	453	1,326	4,551,297	8	5				
415	New legislation for appellation d'origine wines Nuova disciplina delle Denominazioni d'Orgine dei Vini	1992	92,590	190,852	0	7	3	G2			
416	Institution of the Regional Park of Colli Euganei Norme per l'Istituzione del Parco Regionale dei Colli Euganei	1989	n.a.	n.a.	n.a.	2	5				
417	Initiatives under the programme 2052/88 Objective 5b Programma di attuazione del Reg. CEE 2052/88 Obiettivo 5b	1988	295	n.a.	1,172,627	1	4				
418	Framework act on forest and institution of the hydrogeological constraint Riordinamento e riforma della legislazione in materia di boschi e di terreni montani	1923	700,000	7,787,700	0	4	5	G7			
419	Landscape Act (Legge Galasso) Conversione in legge con modificazione del DL 27 giugno 1985 n 312 recante disposizioni urgenti per la tutela delle zone di particolare interesse ambientale	1985	n.a.	n.a.	n.a.	2	6				

Annexe: overview of inventoried CSPs

420	Framework Act on Protected Areas Legge quadro sulle aree protette	1991	n.a.	1,981,289	0	2	5	G5	
421	New Framework Act for Mountainous areas Nuove disposizioni per le zone montane	1994	n.a.	n.a.	n.a.	–	5		
422	Act for wildlife control Nuove norme relative al controllo della fauna selvatica	1993	n.a.	n.a.	n.a.	3	5		
423	Paniere Veneto Paniere Veneto	1988	n.a.	n.a.	n.a.	7	4		X
	Regulation 2078/92 schemes in the Emilia Romagna region (measures 424–441)								
424	A1 Reduced use of agri-inputs or maintenance of reduction A1 Sensibile riduzione dell'impiego di concimi e/o fitofarmaci oppure mantenimento delle riduzioni già effettuate	1993	2,021	26,733	23,196	6	1		2
425	A2 Organic Farming A2 Agricoltura biologica	1993	896	10,240	8,952	6	1		1
426	B1 Agronomic practices for annual crops in plain and hilly areas B1 Pratiche agronomiche da applicare congiuntamente per le colture annuali in pianura ed in collina	1993	399	n.a.	9,296	4	5	G7	2	3
427	B2 Conversion from arable to extensive grassland B2 Gestione di terreni con regime sodivo	1993	n.a.	5,766	1,104,185	4	4			3
428	B3 Agronomic practices for vineyards and orchards in hilly and mountainous areas B3 Pratiche agronomiche da introdurre o mantenere nei vigneti già esistenti e nei frutteti di collina e montagna	1993	8	13	3,099	4	4		1	3
429	C1 Reduction of livestock density in plain areas C1 Riduzione del carico di UBA/ha foraggere nella zona omogenea pianura	1993	n.a.	n.a.	10,329	–	4	G7	3	3
430	C2 Reduction of livestock density in hilly and mountainous areas. C2 Riduzione del carico UBA/ha foraggere nella zona omogenea montagna e collina.	1993	n.a.	n.a.	43,382	–	4		3	3

Italy continued

	Policy	Year of issue	Number of farms	Area (ha)	Funds (Euro)	Ch. 2a	Ch. 2b	Ch. 3	Ch. 4	Ch. 7	Ch. 8
431	D1 Conservation and restoration of elements of the natural environment and/or of the rural landscape D1 Conservazione e/o ripristino di spazi naturali e seminaturali e degli elementi dell'agro-ecosistema e del paesaggio agrario	1993	193	799	1,712,571	7	3			1	3
432	D2 Cultivation of crops for wildlife feeding D2 Effettuazione di coltivazioni a perdere per l'alimentazione della fauna selvatica	1993	38	204	270,623	–	6	G7			3
433	D4 Introduction of cover crops D4 Realizzazione di colture intercalari	1993	n.a.	32	6,714	1	4				3
434	D5 Rearing animal breeds in danger of extinction D5 Specie annuali locali minacciate di estinzione	1993	1,155	10,095	3,021	–	4	G7			3
435	E1 Maintenance of mountainous extensive grazings E1 Cura dei pascoli estensivi di montagna mediante ordinaria manutenzione	1993	n.a.	1,399	143,575	4	2	G8		4	3
436	E2 Maintenance of abandoned coppice woodlands E2 Cura dei boschi cedui abbandonati di collina e montagna	1993	n.a.	1,313	317,105	4	2			4	3
437	F1 Development of natural environments for wildlife F1 Realizzazione di ambienti fisici a carattere unitario, idonei a garantire la sopravvivenza e la riproduzione di fauna e flora selvatiche	1993	n.a.	1,942	1,364,479	2	6	G3		2;3	3
438	F2 Development of natural environments for enhancing rural landscapes and ecosystems F2 Realizzazione di ambienti naturali e seminaturali variamente strutturati con funzioni di collegamento paesaggistico ed ecologico fra elementi territoriali	1993	n.a.	167	130,664	2	6				3
439	F3 Development of natural environment for water protection F3 Realizzazione di ambienti idonei a contribuire alla salvaguardia dei sistemi idrologici	1993	n.a.	9	4,648	2	6			1	3

Annexe: overview of inventoried CSPs

440	G1 Development of footpaths and other recreational structures in protected areas G1 Realizzazione di percorsi obbligati, organizzati nell'ambito dei parchi, riserve naturali dai rispettivi enti di gestione dei piani di fruizione naturalistica, turistico-ambientale e del tempo libero...	1993	n.a.	n.a.	n.a.	5	4	G7	4
441	G2 Development of turistic/recreational structures in rural/historical buildings or near hydraulic works G2 Realizzazione di idonee sistemazioni atte a favorire l'accesso al pubblico od attività culturali e ricreative in prossimità di manufatti idraulici, di edifici di interesse storico o di notevole valore archiettonico inseriti in ambienti naturali	1993	4	6	2,066	5	4	**G7**	4
442	Rural tourism Misura 6.5 Turismo Rurale (Sottoprogramma Turismo)	1995	n.a.	n.a.	3,000,000	5	3	G7	
443	Conservation of protected natural areas Sottomisura 7.3.9 Tutela e conservazione di aree naturali protette (Misura 7.3 Ambiente)	1995	n.a.	n.a.	n.a.	3	6	G7	
444	Promotion of typical regional food products Misura 4.3.2 Qualificazione, valorizzazione, e promozione dei prodotti agroalimentari tipici di qualità (Sottoasse 4.3 Servizi di Sviluppo e Divulgazione)	1995	n.a.	n.a.	n.a.	–	4	**G7**	
445	Demonstration farms for organic agriculture Azione dimostrativa per la messa a punto e la diffusione dei metodi di produzione biologica	1995	n.a.	n.a.	n.a.	–	4	**G2**	
446	Development of Extension services for agr-food system Misura 4.3.6 Assistenza al programma e supporto tecnico-scientifico al sistema agro-alimentare (Sottoasse 4.3 Servizi di sviluppo e Divulgazione)	1995	n.a.	n.a.	n.a.	–	7	G7	
447	Agritourism Misura 4.2.1 Agriturismo (Sottoasse 4.2 Sviluppo rurale)	1995	158	n.a.	n.a.	–	5	**G7**	

Italy continued

	Policy	Year of issue	Number of farms	Area (ha)	Funds (Euro)	Ch. 2a	Ch. 2b	Ch. 3	Ch. 4	Ch. 7	Ch. 8
448	Enhancement of protected areas Sottomisura 7.3.10 Valorizzazione e fruizione di aree naturali protette (Misura 7.3. Ambiente)	1995	n.a.	n.a.	n.a.	5	5	**G5**			
449	Protection of rural heritage and buildings Misura 4.2.3 Ristrutturazione e valorizzazione del patrimonio rurale	1995	n.a.	n.a.	n.a.	–	2				
450	Woodland maintenance Misura 4.2.4. Miglioramento e cura dei boschi	1995	n.a.	n.a.	n.a.	3	2	G7		4	
451	Protection and valorisation of typical rural buildings (trulli di Alberobello) Tutela dell'ambiente naturale caratteristico e dei monumenti pugliesi e valorizzazione, salvaguardia e destinazione d'uso dei trulli di Alberobello	1979	n.a.	n.a.	n.a.	–	5	**G7**			
	Regulation 2078/92 schemes in the Puglia region (measures 452–458)										
452	A2 Organic agriculture A 2 Introduzione o mantenimento di metodi dell'agricoltura biologica	1995	524	14,256	5,328,169	1	2			1	
453	B3 Conversion to extensive production methods B3 Riconversione dei seminativi in pascolo	1995	1	18	n.a.	6	1			1	
454	C Reduction of stocking density C Riduzione della densità del patrimonio ovino o bovino per unità di superfici a foraggere	1995	1	n.a.	8,117	1	5			3	
455	D2 Maintenance of natural environment and landscape D2 Cura dello spazio naturale e del paesaggio	1995	177	6,660	698,682	1	4			1	
456	D3 Rearing of breeds in danger of extinction D3 Allevamento di specie animali in pericolo di estinzione	1995	92	452	74,620	3	4				
457	E-Maintenance of abandoned forest land E Cura dei terreni agricoli o forestali abbandonati	1995	18	413	56,114	4	2			1;4	

Annexe: overview of inventoried CSPs

No.	Name	Year						
458	F Twenty-years set-aside F Ritiro di seminativi dalla produzione per almeno 20 anni, nella prospettiva di un loro utilizzo per scopi di carattere ambientale'	1995	2	41	31,214	2	6	1
459	Contribution to meet damages caused by wildlife Contributi per danni causati da specie animali di notevole interesse scientifico - Modificata dalla LR105 del dicembre 1994	1974	n.a.	n.a.	49,507	—	5	
460	Legislation for mushroom picking Normativa per disciplinare la raccolta dei prodotti del sottobosco e per la salvaguardia dell'ambiente naturale	1982	n.a.	n.a.	n.a.	—	4	
461	Ban for felling olive trees Divieto di abbattimento di alberi di olivo Modificata dalla LN144 del febbraio 1951	1945	n.a.	n.a.	n.a.	3	4	
462	Protection of chestnut woodlands Provvedimenti per la tutela dei castagneti e per il controllo delle fabbriche per la produzione del tannino dal legno di castagno	1931	n.a.	n.a.	n.a.	—	5	
463	Protection of trees and shrubs in the area of the Commune of Florence Regolamento per la tutela del patrimonio arboreo ed arbustivo del Comune di Firenze	1991	n.a.	n.a.	n.a.	—	6	

Regulation 2078/92 schemes in the Toscana region (measures 464–471)

No.	Name	Year						
464	A1 reduction and maintenance of the use of chemical inputs A1 Riduzione e mantenimento della riduzione di concimi e fitofarmaci	1994	n.a.	111,907	55,666,965	6	1	2
465	A2 Organic agriculture A2 Introduzione e mantenimento dei metodi dell'agricoltura biologica	1994	n.a.	10,013	3,409,322	6	1	1
466	B1 Introduction of extensive methods of cultvation B1 Estensivizzazione delle produzioni vegetali con mezzi diversi dalla riduzione dell'impiego dei concimi chimici e fitofarmaci	1994	n.a.	2,320	520,972	4	4	

Italy continued

	Policy	Year of issue	Number of farms	Area (ha)	Funds (Euro)	Ch. 2a	Ch. 2b	Ch. 3	Ch. 4	Ch. 7	Ch. 8
467	C1 Reduction of stocking density C1 Riduzione della densità del patrimonio bovino ed ovino per unità di superficie foraggera	1994	n.a.	0	115,822	6	4			3	
468	D1 Methods for the protection of environment and the rural landscape D1 Impiego di altri metodi di produzione compatibili con le esigenze di tutela dell'ambiente e delle risorse naturali del paesaggio	1994	n.a.	1,972	532,380	1	3			1;2	
469	D2 Rearing of breeds in danger of extinction D2 Allevamento di specie e razze animali locali minacciate di estinzione	1994	n.a.	n.a.	406,485	3	4				
470	D3 Cultivation of local vegetable species in danger of extinction D3 Coltura e moltiplicazione dei vegetali adatti alle condizioni locali e minacciati di erosione genetica	1994	n.a.	707	223,868	3	5				
471	F1 Twenty years set-aside F1 Ritiro dei seminativi dalla produzione per almeno 20 anni	1994	n.a.	929	946,891	2	6				

Annexe: overview of inventoried CSPs 225

Sweden

	Policy	Year of issue	Number of farms	Area (ha)	Funds (Euro)	Ch. 2a	Ch. 2b	Ch. 3	Ch. 4	Ch. 7	Ch. 8
501	Conservation of biodiversity and cultural heritage values in mowed meadows Miljöstöd för bevarande av biologisk mångfald och kulturmiljövärden i slåtterängar	1996	1,956	4,893	4,199,965	–	4	G1	X	4	1
502	Conservation of biodiversity and cultural heritage values in semi-natural grazing lands Miljöstöd för bevarande av biologisk mångfald och kulturmiljövärden i betesmarker	1996	15,296	198,744	83,994,149	–	4	G1	X	4	1
503	Conservation of areas with biologically rich habitats and valuable cultural heritage environments Miljöstöd för bevarande av värdefulla natur- och kulturmiljöer	1996	11,488	614,319	89,797,827	–	4	G1	X	4	1
504	Maintenance of an open landscape in northern Sweden and in the forest regions Miljöstöd för bevarande av ett öppet odlingslandskap	1996	41,021	775,186	273,703,165	–	4	G1	X		1
505	Restoration and establishment of wetlands and ponds on arable land Miljöstöd för anläggning eller återställande av våtmarker eller småvatten	1996	311	1,242	1,406,866	–	5	G1			1
506	Establishment of permanent grassland to prevent nutrient leakage and erosion Miljöstöd för anläggning av extensiv vall och skyddszoner	1996	1,319	2,600	1,476,903	–	4	G1			
507	Promotion of catch crops Miljöstöd för fånggrödor	1996	844	8,381	1,252,550	1	4	G6			1
508	Conservation of local breeds threatened by extinction Miljöstöd för bevarande av utrotningshotade husdjursraser	1996	1,116	n.a.	1,225,199	3	4	G5			4
509	Traditional cultivation of local varieties of brown beans (*phaseolus vulgaris*) on the island of Öland Miljöstöd för traditionell odling av bruna bönor på Öland	1996	139	974	1,009,495	–	4				
510	Promotion of organic production Miljöstöd för ekologisk odling	1995	15,137	256,704	91,576,540	–	4	G2		1	1

Sweden

	Policy	Year of issue	Number of farms	Area (ha)	Funds (Euro)	Ch. 2a	Ch. 2b	Ch. 3	Ch. 4	Ch. 7	Ch. 8
511	Directive regarding acreage compensation for energy crops	n.a.	n.a.	n.a.	n.a.	1	7				
512	Acreage compensation for forage production	1997	46,316	714,262	95,975,861	1	7				
513	Environmental charges on fertilizers Miljöskatt på handelsgödselkväve	1984	n.a.	n.a.	n.a.	6	7	G7			
514	Environmental charges on biocides Miljöavgift på bekämpningsmedel i jordbruket	1984	n.a.	n.a.	n.a.	6	7				

United Kingdom

	Policy	Year of issue	Number of farms	Area (ha)	Funds (Euro)	Ch. 2a	Ch. 2b	Ch. 3	Ch. 4	Ch. 7	Ch. 8
301	Nitrate Sensitive Areas Scheme	1995	359	19,671	17,512,393	6	1	G4	X	1;2;3	3
302	Inheritance tax (Capital taxation and heritage landscapes)	1986	189	106,000	7,568,018	1	3				
303	Sites of Special Scientific Interest	1981	2,670	106,773	50,082,296	3	4	G1	X		1
304	The Moorland Scheme	1995	20	4,491	1,872,328	2	5	G1	X	1;2;3;4	1
305	The Countryside Access Scheme	1994	110	1,492	1,000,492	5	6	G7	X		2
306	The Habitat Scheme	1994	388	6,698	7,928,255	3	3	G7	X	1;3	2
307	Environmentally Sensitive areas Scheme—Composite questionnaire for all English ESAS	1987	7,792	426,683	243,296,628	6	1	**G1**	X	1;2;3;4	2
313	Organic Aid scheme (England)	1994	101	4,673	2,141,749	6	1	**G2**	X	1;3	1
314	Countryside Stewardship Scheme	1991	5,304	96,227	67,170,616	2	4	G1	X	4	2
315	National Parks and Access to the Countryside Act	1949	44	27,821	806,974	5	3		X		1
316	Farm Woodland Premium Scheme	1992	2,373	7,336	16,523	–	7	G1			1

Index

administrative
   authorities 38
   costs 69–70, 76, 79
   structure 15–16
   time requirements 80
agenda 2000 7–8, 30, 87, 132, 153, 161, 186–187, 190–191
agreement on agriculture (AOA) 157,158, 161, 166
agricultural intensity 10
agro-chemicals 128, 131, 194
AOC [appellation d'origine controlée] 32, 58, 61
asymmetry of information 67, 69–70, 76, 183
attitude
   analysis model 98
   differences 101
   measurement 92–93

beef extensification scheme 114, 132–133
benchmarks (also boundaries, reference points) 21, 24, 26–28, 164, 181
beneficiary pays approach [principle] 31, 40, 44, 187–188
biodiversity
   economic nature 24–27
   objective 32, 34, 171, 178
   private transaction costs 110
Birds directive 6
blue box measure 162
book
   structure 5
   outline 16
boundaries (see benchmarks)
by-product (also joint product) 2, 23

club good 23, 188
cluster[ing]
   for trade distortion analysis 172–174
   for typology 58–59, 62–64
collective agreements 86
commoditisation 31
compatibility with policy principles 50–54, 66, 175–178, 188
compensatory payments (also remuneration) 17, 38–39, 54, 60–64, 74, 77, 123, 189–190

competitive relationship 23
complementarity relationship 23, 64
compliance costs 70
consumer concerns 2
control 38, 58, 184
covenants 31, 36
cross compliance 31, 184, 190
CSP (Countryside Stewardship Policies)
   administration 37, 79
   attributes 35
   authorities 38
   classifications 64–65, 137, 170
   compensation 35
   definition 3, 21, 32
   expenditures 39, 76–78
   funds 39
   objectives 34, 136, 178
   target areas 39, 182
   technical aspects 42
   typology 55–65

decoupling of support 53, 162
degradation
   of biodiversity 11
   of land use 9
   of water resources 10
designation 182
development of ERGSs 22, 31, 117
direct payments
   influence 125–126, 131–132,
   instrument 135–137,
   model results 146–153
   role of 140, 144,
Dobris assessment 11
dual value of land 125

EAGGF 136, 142, 153
econometric models 110,115
economic mechanisms 3, 30
EDM (Equilibrium Displacement Model)
   method 144–147
   parameters 148–150
   results 151–152

effectiveness
  cost effectiveness 68–70, 86, 182, 184–185, 189
  criteria (indicators) 5, 58
  definition 50
  of output reduction 131–133, 174–175
efficiency
  criteria 52
  definition 50
  discussion 180
  of costs (cost efficiency) 52, 78–85, 181–184, 189
eligibility 47, 50, 52, 182, 189
enforceability 47, 50, 52
enforcement costs 17
environmental attitudes (see attitudes)
environmental indicators 48
environmental (recreational) bad and dis-service (EBD or ERBD) 22–23
environmental (recreational) good and service (EGS or ERGS) 22–23
Equilibrium Displacement Model (see EDM)
equity (also fairness) 47, 52, 59, 149
EU funds 17, 39, 187
EU environmental policies 6–8
EU regulation 2078/92 1, 7, 17
excludability 22–23, 28, 31
experience of farmers 103, 105, 109
externalities
  boundaries 27
  definition 22
  discussion 54, 180–181
  technical aspects 42
  typology 60–62, 65

factor analysis 55
fairness (also equity) 53, 187
farm behaviour
  model 90–92
  questionnaire 93–94
  data set 94–95
  hypotheses 96
  results 98–109, 179
farm income
  influence on attitudes 106–107
  effects 124, 130–132, 152–153
farmer-pays-mechanism 187
flexibility 50

GATT (also WTO) 136, 158, 161
good agricultural practices 28, 45, 141
green box measure 136, 162, 166

horizontal measure [policy] 8, 141, 148, 153, 183
household utility function 90–92

incentives (see compensatory payments)
incidental by-products 24, 58
income and environment 2,3
  elasticity 2,3
  support [direct] 8
input restrictions 139
institutional aspects 1, 15–16, 37, 73, 84, 184
instruments (also policy tools) 28–32, 36, 137–138, 183
internalisation 22, 28, 30, 163
international trade (also trade distortion) 66, 136, 158, 188

joint product[ion] 23, 86

kernel density function 151

label[ling] (also AOC) 32, 34, 36–37, 50, 61, 65–66, 170, 187–188
land abandonment 2, 30, 113, 126, 130–132, 138–140, 186
land values 125
legal boundaries (also benchmarks) 26, 28
leisure 90, 148, 150
likert scale 89, 92, 101
local public goods 47, 60–61, 65
logit model 89, 96, 108
long term contracts 110

mandatory measures 28, 35, 39
marginal land 113, 125, 130–131, 138, 144
market
  access 161
  creation 31
  distortion (also trade distortion) 180–181
  effects 41, 141, 179, 186–187
  mechanisms 2, 163, 184
marketing of EGS 31, 40–41, 70
mixed goods 23
modulation 181, 188
multi-functional 25, 182
multiple correspondence analysis 47, 57
multiple regressions model 96
multipurpose farming and forestry 25

negative external effect (see externalities)
nitrate directive 6
non-participation
  reasons for 99–100
  model 103–106

Oberösterreich 18, 120–122, 124–129
off-farm work 90

opportunity cost 68–70, 80–81, 84, 96, 110, 132–133, 137, 141, 178, 184
OPUL
    description 122–123
    effects 124–130
organic farming 58, 76, 82, 123, 133
output effects (also market effects)
    discussion 186, 189
    Equilibrium Displacement Model results 149–152
    regional model 116–118
    regional model results 129–135
    survey results 143–144
over-compensation 53–54, 98, 131, 179, 183

participation
    reasons of 98–99
    model 103–106
PMP
    model 116–118
    study regions 120
    income effects 124
    shadow values 125
    results 124–133
policy
    acceptability 50, 52, 187
    administration 15–16, 76, 183
    design 85
    indicators 13, 48, 51–53
    instruments (see tools)
    outcomes 178, 181–183
    targeting 166–167, 170–175, 182
    tools 28–32, 36, 137–138, 183
    uptake 182
political acceptability 50, 52, 187
polluter pays principle (PPP) 3, 30, 53–54, 60–63, 159–160, 187–188
pollution 6, 10, 11
positive external effects (see externalities)
positive mathematical programming (see PMP)
positive quadratic programming (see PMP)
premium 123, 138–139, 148
prescription 72, 76, 81, 85, 159–160, 180–182, 185, 188
prime à l'herbe 120, 130, 131, 138
private goods 21–23, 188
private transaction costs 73, 76, 84, 86, 91, 96–97, 179, 183
producer support estimate (PSE) 185, 164–165
product quality 61
production effects (also output effects) 113, 115, 124, 126, 130–131
production possibility curve (PPC) [or PPF production possibility frontier] 23–26, 25, 43–44

property rights 2–3, 8, 16, 21, 23, 26, 28, 31, 35–36, 44, 52–53, 69–70, 137, 140, 146, 152, 182, 187–191
protected area 14
public exchequer costs 68, 74
public good 22–23, 28, 32, 62–69, 137, 152, 178, 180, 182, 187–189
pure private goods 23
pure public goods 22, 62, 65

quasi-decoupled measures 162
quasi-markets for environmental goods and services 137, 152
questionnaire 32, 93–94

recommendations 190
reference points (also benchmarks) 3, 21, 26–28, 178, 181
remuneration (also compensatory payments) 2, 17, 32, 35, 37–38, 111, 180, 183–184, 189
rent(s) 31, 52, 73, 125, 141, 148, 153, 179, 182, 183, 186
rivalry 22–23, 28
rural development 7–8, 29, 31, 45, 177, 184

sampling procedures 96, 98
set up costs 79
set-aside 78, 127–128
state-pays-approach 30–31, 84, 187
stewardship definition 2
substitution 25
sustainability 50, 52, 54, 61–62, 182, 189–190

tailoring of policies 157, 174–175, 180, 187
targetability 51–52
targeting 83, 86, 92, 110–111, 157, 163, 171, 174–175, 180–183, 187–189
taxes 3, 28, 30, 73, 188
taxpayer 50, 164–165, 187, 189
technologies 54, 58, 113, 115, 159, 179, 182
trade
    criteria 166–167
    distortion 4, 9, 157–158, 161, 164–168, 175, 180, 184, 187
    effect 47, 49, 160, 163, 177, 186
    evidence 160
    measurement 164
    policies 4, 161
    qualitative results 168–169
    quantitative results 170–175, 186
    theoretical considerations 158–160

trade-off
  between F&F and EGS 3, 41–43
  between compensations and transactions costs 70, 83, 86, 181, 185, 190
  between objectives 34, 153, 190
transactions costs
  and commodity regimes 78
  definition 69–70
  economics of scale 81
  estimates 76–78
  influencing factors 83, 184–185
  level 52–53
  measurement 74–75
  private 73–76, 96–97
  reduction 84, 184–185
  relative importance 79, 178
  trends in time 79
  typology 71–72
transparency
  of costs 83, 87
  of policies 50, 52, 59, 167, 171–174

typology of CSPs
  methodology 55–58
  factors 58–59
  results 59–64

uptake
  factors 89–91, 98–99, 103–109, 179
  increase 182–183
  level 17, 39, 61–66, 101
Uruguay Round Agreement (see agreement on agriculture)

vertical measures 8, 25, 61, 64, 183
voluntary measures 21, 28–29, 35–36, 39, 182

water protection 51, 53, 59, 138

zoning 52–54, 58–61, 181–182, 189